D1560682

CHICAGO PUBLIC LIBRARY
HAROLD WASHINGTON LIBRARY CENTER
R0017266023

cop 1
THE CHICAGO
PUBLIC LIBRARY
C
P
L

Desk Book For Construction Superintendents

King Royer, P. E. has held jobs in construction firms from journeyman carpenter to secretary of a firm grossing a million dollars a month. He has been an estimator, job superintendent, project manager, and chief engineer for firms in half a dozen states. He has also been a supervising engineer for several branches of the U.S. government and a professor of building construction at the University of Florida.

SECOND EDITION

Desk Book For Construction Superintendents

KING ROYER

Building Construction Department
University of Florida

Prentice-Hall, Inc., *Englewood Cliffs, N.J. 07632*

Library of Congress Cataloging in Publication Data

Royer, King.
Desk book for construction superintendents.

Bibliography: p.
Includes index.
1. Building–Superintendence–Handbooks, manuals, etc. I. Title.
TH438.R68 1980 658.5 79-17032
ISBN 0-13-202028-9

Editorial/production supervision
and interior design by Lori Opre
Cover design by 20/20
Manufacturing buyer: Gordon Osbourne

Printed in the United States of America

10 9 8 7 6 5 4 3 2 1

Prentice-Hall International, Inc., *London*
Prentice-Hall of Australia Pty. Limited, *Sydney*
Prentice-Hall of Canada, Ltd., *Toronto*
Prentice-Hall of India Private Limited, *New Delhi*
Prentice-Hall of Japan, Inc., *Tokyo*
Prentice-Hall of Southeast Asia Pte. Ltd., *Singapore*
Whitehall Books Limited, *Wellington, New Zealand*

Dedicated to

Frank B. Gilbreth
who in 1905 was familiar
with all ideas in this book.

Contents

Preface

In recent years, management is considered to be people who sit in offices and make decisions, like judges, based on information brought to them. The large firms in construction, as in other industries, attempt to organize their work in this way, but the construction manager fits this assumption so poorly that academic management authors ignore him entirely.

The construction superintendent is one of many people in the construction industry who must make decisions, not coldly and carefully, but in a few minutes and with little information. Managing is part of the work of each person, and there is little demand for a person with no other duties than to "manage" on the basis of information gathered for him. This book is intended to add to the management store of information needed by the job superintendent.

Seventy years ago, John P. Slack wrote: "The direction of a large body of men of various degrees of intelligence, working in different localities, some perhaps far removed from headquarters, is a proposition requiring experience, brains, and a highly specialized training in the art of handling men. The large contractor executing many contracts simultaneously has this proposition to face, and the problems which must be solved are many and puzzling."

The superintendent's problems are now even more complicated, and the key to a contractor's success is the ability of his job superintendents. It is hoped that this book will help to improve the contractor's control of costs and the ability of superintendents, both by its use by working supervisors and by students who will be the future supervisors.

This book is directed to the particular duties of the superintendent to prevent the narrative from expanding into a general description of what happens on a construction job. What happens is not nearly so important to the participants as *what makes it happen.* The small contractor, the developer, and accountant, however, will also find considerable information of value to them, and the architect and engineer will gain a better appreciation of the superintendents with whom they work daily. The small contractor will consider many of the duties described beyond the authority of the job superintendent; on his work, the superintendent has limited responsibilities. But the practices described offer, to contractor and superintendent alike, methods to make operation of large jobs simpler and above all, *more profitable.*

The Glossary includes words that are generally well known, but may not be familiar to students; words that have a formal construction meaning in addition to a common meaning, and a few words that the author has his own particular comments on, but may not be of interest to the student.

The author will appreciate any comments and criticisms that will eventually reach him if addressed as below.

KING ROYER

1237 S. W. 9th Road
Gainesville, Fla. 32601

INTRODUCTION

1

This book is written for the trade foreman or student. It is expected that the reader knows or will learn elsewhere the technical knowledge of his own and associated trades. You will have to learn from experience countless details of supervision, layout, and application.

This book won't teach you your trade. There will be parts of this book that are entirely unnecessary for you because you know them. There will be portions with which you will entirely disagree. This is because each person's attitudes—yours and mine—are based on his own experience. There is nothing new in this book; most contractors are familiar with practically all the material in it. Its purpose is to help you to understand why your boss tells you to do things as he does. Often superintendents are not told why things are done and make ridiculous mistakes on this account. If, on the other hand, you are considering going into business for yourself, you will need to know as much as a superintendent, and more.

Generally, in this book it is assumed that you are a superintendent on a $1 to $3 million building project, that your office help consists of a clerk and possibly an assistant who is the layout engineer, and that you are responsible to a contractor in a home office in another municipality. The contractor has an estimator-office engineer and a bookkeeper. It is also assumed that you are young and have little, if any, experience. Your experience with labor unions and suppliers is based on work in another area.

For other situations, you may make appropriate changes in the descriptions of positions—*project manager* rather than *contractor,* for example,—and many items shown here to be done by you are often handled by an office engineer.

There is very little that is new in this book. Although the writer's opinions are usually obvious, an attempt is made to show various methods so that you may choose the one that suits your job. As the person in the middle, you usually have very little choice in your organization, jobs, type of completed work, accounting procedures, or policy toward labor unions, and your subcontractors and architect are selected by others. You must be prepared to do the best you can—to bring together many often mutually antagonistic elements into a completed building. I will not add to your troubles by imposing yet another series of recommendations or rules.

Although written for the superintendent and student, this book should be equally useful for those owners, lacking a trade background, who are supervising their own construction and who depend on the superintendent they hire for managerial ability and knowledge of technical matters.

The few office management and cost accounting methods described are examples of good practice and common usage. The specific forms used and procedures followed will vary from one firm to another. If you understand what information is wanted, you can pick up details of reports in a few minutes. No effort is made in this book to consider legal problems, except as they affect the job superintendent.

There is a fundamental difference between the duties of a superintendent and those of a foreman. A superintendent is responsible for *all* the trades, and the one who complains that a mistake is the "plumber's fault" is merely passing the buck for his own ignorance. A superintendent is usually an ex-carpenter foreman, because carpenters have the most varied trade knowledge and come into contact with the work of nearly all the other trades. When acting as superintendent, such a person must wean himself from his own trade and learn others. He must learn to read plans, to spot conflicts between trades, to estimate time, to judge workmanship, and to give as much care and attention to the other trades—particularly the mechanical trades—as to his own. A great deal of the delay and error in construction operations is caused by a superintendent's failure to watch *all* the trades.

The superintendent must take the initiative for all operations. He must remember that there is a hole under the hydraulic elevator, that the steamfitter has sleeves in a masonry wall, that there are anchor bolts in the floor. He must see that the shop drawings fit the job as built, and that the sprinkler fitter is notified when a foundation wall is to be poured. He should be sure that the temporary elevator is not placed where the drain lines enter the building, and should not permit temporary drainage grading that will prevent installation of piping outside the building. He is ultimately responsible for errors in the plans. If the plans were properly prepared, if all subcontractors did their work properly, and if the general contractor's personnel were fully competent, the superintendent's job would be unnecessary. It is unlikely, however, that this happy coincidence of events will happen in the near future.

As the contractor's representative, the superintendent must be conscious of the public at all times. At best, his work will often be resented by others, and his failure to foresee and correct dangerous conditions can cause serious damage to the contractor.

Just What Is Your Job?

A superintendent being hired was told by a contractor, "I want someone to take over the whole job for me, to take care of all my worries."

"That's some job," said the applicant, "What's the pay?"

"Five hundred dollars a week."

"That sounds like a lot of money. Where is the money coming from?"

"That will be your first worry."

2.1 DUTIES OF THE JOB SUPERINTENDENT

If you have been working with one firm, on one kind of job, or in one community, you may believe you know what a superintendent should or should not do. But you may be in for some surprises.

You are familiar with superintendents who are visited daily by the contractor, for example, and now you find that some superintendents never receive a visit from the contractor until the job is completed. Or, you may be working on a concrete floor when the superintendent looks at it and says, "Tear it out–it's wrong!" You point out that it is in accordance with the drawing, but he retorts, "If the drawing shows it that way, the drawing is wrong." Or, you may be installing a septic tank and hit a spring. The superintendent takes one look and says, "Cover it up. We'll put the septic tank above ground."

A few such occurrences confuse you; you wonder where the superintendent is getting his information. These are unusual examples, but ones that might well occur. In each case the superintendent, beyond merely supervising construction, had authority to change the plans as necessary.

A superintendent's authority is not the same on each job. Some superintendents—the best ones—require that all people on the job look to them as the contractor's representative, and they would quit if the contractor interfered. Others are merely carpenter foremen who relay messages to the contractor. Many specifications require that the superintendent be authorized to represent the contractor in all matters regarding the job, but few contractors are willing to delegate this authority. On owner-built jobs, the superintendent, when dealing with the architect, may represent the owner, so the superintendent actually supervises the architect rather than the other way around. On jobs not requiring a full-time supervisor, the superintendent may merely be one of the workers, to direct the work and represent the contractor.

2.2 TRADE SUPERINTENDENTS

A superintendent is a contractor's or subcontractor's representative; that is, each trade subcontractor, such as the plumbers, electricians, roofers, or others, has its own representative, who may be called foreman in smaller jobs or superintendent on larger ones. The general contractor's superintendent is designated as the general superintendent; the others are plumbing superintendent, electrical superintendent, and so on. In a job so large that a single superintendent cannot oversee all of it personally, he has assistants, usually designated as area superintendents. These people are responsible for a particular building or other part of a job.

If a general contractor hires all workers directly, instead of subcontracting part of the work to others, he will still need trade superintendents. Sometimes *all* trades have their own superintendent, including the carpenters; in such cases, the general and area superintendents do not have hiring and firing authority, and may be less experienced than they would need to be otherwise. Area superintendents in this case are often recent college graduates in construction or engineering.

2.3 SUPERINTENDENT'S AUTHORITY

The authority given a superintendent depends on his status in the company (that is, how much he is trusted), the size of the job, and the type of company. Older and larger firms often have duties more carefully defined, and their superintendents therefore have less authority.

The basic authority of the superintendent may include some or all of the following:

1. To hire workers. On open-shop work, the workers may be people he or his foremen know. On union work, the superintendent is more likely to accept workers in a union-designated order.
2. To designate and discharge foremen; or if the superintendent is also a foreman, to assign and discharge workers.
3. To lay out and direct the work, to interpret plans and specifications, and to receive instructions from the architect and engineer —to whom he represents the contractor.

In addition, the superintendent may have additional authority:

4. To decide the size of the work force.
5. To change from union to open-shop workers, or vice versa. This authority is rarely given.
6. To decide rates of pay, to authorize overtime work, and to pay bonuses.
7. To give weekly pay guarantees to individual workers. Some workers, and often foremen, are not normally paid for days of bad weather or when work is delayed for other reasons.
8. To buy materials.
9. To cancel office purchase orders from one supplier to another.
10. To approve material and subcontractor payments.
11. To cancel subcontracts.
12. To estimate extra work and to provide the architect or engineer with estimates.

You will not have your authority defined by the contractor in an hour's time. In fact, it may take 20 years. Your authority may be specified by definite written instructions or, more often, by a series of verbal orders over a period of time.

Most contractors aren't definite: if they give written instructions, a good superintendent's authority may be too greatly limited, while a poor superintendent may be given too much authority. Also, self-confidence is an important factor in assessing a superintendent. Some employers automatically say *no* when a superintendent asks for authority and yet will approve his actions if he acts without authority. This may be because the employer believes that if a person asks for authority, he isn't sure of himself—if he knew there were but one decision to make, he'd make it. *Do your job right, and don't worry about "giving the boss an excuse to fire you."* If the job doesn't go right, he'll disregard everything else. If it goes well, you'll get the credit anyway. One employer who was asked to approve an expenditure said, "Of course I'm against it. I'm *always* against spending money. The question is: What else can we do?"

So, if you don't know if you should assume responsibility, don't.

When you ask the boss, he'll let you know whether you should have asked, usually a question such as: "Can't you handle these matters yourself?" And when he chews you out for doing something you shouldn't have done, listen carefully! It makes a great deal of difference whether he's telling you that your mistake was the *actual decision you made*, or that it was *making it without asking him*. There are some employers who have it both ways: when you ask, they say you shouldn't need to, but when you make a mistake, they say you should have asked. But this is logical; if you make a wrong decision, this is proof you weren't qualified to make the decision in the first place! Contractors are seldom concerned with explaining who does what and in being consistent about it—they just want the job done cheaply, and if it isn't, they are unhappy.

2.4 STARTING THE JOB

When you start a job, you will find that you have to take the initiative. If you haven't the authority to purchase, you still must remind someone that it has to be done and follow up afterward to see that everything is bought. Since the superintendent must remember items which are not on the plans anyway, contractors often allow superintendents to purchase such items, even though the main office may do all other purchasing. Such items may include the job office, formwork materials, and arrangements for temporary water, telephone, and electric service.

Good telephone service is important. It should not be necessary to have a clerk just to answer the telephone. A locked-box waterproof phone can be installed anywhere on the job—this is the kind of telephone taxi companies and police use. It can be installed on a post, and moved or removed when necessary. The telephone company doesn't expect to make money on installations that remain for a short time, so their charges for these installations are very moderate. If your telephone is in a job office, locate your desk so that you may watch the work—especially incoming traffic—while you are at your desk. Get plenty of pay telephones—usually they not only do not cost anything, but may even earn your company a rental percentage. This will keep workers off your telephone. If the job warrants it, provide subcontractors with a telephone, otherwise, make them use a pay phone. This means that you will have to keep a supply of change in your office; you won't be reimbursed for many of the coins you give out, but it keeps your own line clear for calls.

Small jobs may require only a single telephone line with a few *extensions;* larger ones may have several lines for the general contractor and others for subcontractors. Normally, all telephones on the job should be extensions. Bells may be installed on some or all of them, and buzzers may be used so that the job clerk may call the superintendent on another phone. If your home is reasonably close—within 2 or 3 miles—you may want to get an extension to the office phone at home. An extension can be placed miles away from the main telephone for a comparatively low charge; in fact, an extension a mile away may cost less then a separate phone. A telephone in the superintendent's automobile is useful for a large job and provides a direct hookup to the outside. If the office is not going to be manned continuously, arrange for an answering service or recording attachment to take calls. Be sure that you have all your foremen's home phone numbers for possible emergency use.

There are two attachments available that will permit you to talk on the telephone while leaving both hands free: a speakerphone and an operator's headset. The speakerphone leaves you entirely free but is harder to understand especially if there is much outside noise or you are in a large room. The headset is more awkward to use when a single call is being made, but it is very convenient for continuous calls. If you have to spend much of your time on the telephone, it will be worthwhile to have one that is comfortable and will allow you to use waiting time profitably, such as in reading or writing.

Food-dispensing arrangements are available on large jobs. If you don't get the coffee trucks under control, they will be everywhere at all times, particularly during the lunch period. Workers trying to beat the rush will quit earlier and earlier if the truck isn't kept off the job until lunch time. Bottle dispensers are often forbidden to prevent the scattering of bottles throughout the work area. Paper cups may also be scattered around, even though safety rules and good practice require a garbage can for cups at the water can.

2.5 THE PROJECT MANAGER

The title *project manager* is used by architects, engineers, and subcontractors, as well as by general contractors, to indicate a person whose duties relate to buying and contracts, or who has responsibility for a particular job. You are interested primarily in the contractor's project manager.

A contractor may have a project manager on the site, in the office, or both. If in the office, he will be responsible for all the details mentioned here as being handled in the home office. Usually, he does not supervise the superintendent in the execution of the work.

A field project manager is a person on the site, usually considered to represent the contractor and supervise the superintendent. However, he usually has no trade background and rarely interferes with the superintendent in the execution of the work. Often his pay is less than the superintendent's pay. He will make purchases and is responsible for materials, payments, and other matters not directly related to supervision of the work force. Sometimes he is subordinate to the superintendent and is called the project engineer. The titles vary considerably from one firm to another.

2.6 LICENSES AND PERMITS

If your firm has not been working in a particular state, a corporate permit must be obtained from the state capital. Other licenses may be required by more than one public body; for example, the state, county, and city may each require a license to operate. In addition, you will need a building permit from the city or county and additional permits for each mechanical trade. The mechanical permits are usually obtained by the various subcontractors. Remember that the corporate permit is just a permission for the corporation to be in business; it is not a substitute for local licenses, which may be required for all businesses, and neither of these replaces the contractor's license, which is required to engage in the specific business of contracting. In short, you may need:

1. *Corporate permission*—in your own state, the corporate charter covers this.
2. *Business permit*—which may be state, county, city, or all three, and is usually a tax on the operations of all businesses, whether or not they are corporations. You may not need a business permit for a job in a limited area, such as a city, unless you have a permanent office there; laws vary.
3. *Contractor's license*—this may be a state, county, or city requirement, demonstrating technical and financial ability, in addition to the business permit. The public body granting the license may not be the same one that grants the business permit. For example, you may need a state contractor's license before you may buy a city business permit. In some cases, only an individual, not a corporation, may be granted a license; in such cases, this individual may be required to sign the request for the building permit. Licenses usually require an examination, and may not be granted for several months.
4. *General building permit*—from the city or county (or other agencies).
5. *Specialty contractors' permits*—obtained by the electrician and plumber for their work.
6. *Special permits for any use of public property*—for example, if a sidewalk is to be occupied for construction use, a special permit is often required. Special permits may be required for entrances into

highways, for drainage into streets or public sewers, and for burning trash.

7. *Environmental permits*—these have become so complex that it sometimes takes much longer for permits than for design and construction. You are obliged to assume that others have obtained such permits before a building permit is given.

Generally, the U.S. government is exempt from permits, although when other agencies dispute this, it is easier to get permits than to argue the matter. County and city construction may still be subject to their own agencies' permits. Do not assume that a building permit covers other matters, as various departments of government are involved. In addition, there may be special permits because of the nature of the building, such as approval by a state hotel commission of a place of public accommodation and a fire marshal's approval of a place of public assembly or of hazardous use. These special permits should be obtained by the architect before bids are obtained, but the superintendent should check to see that it has been done to the extent that he is aware of the requirements. The air is also public property; any projections into air lanes must be approved and lighted in accordance with federal regulations. You must inquire regarding air safety lights on temporary structures, such as tower cranes, which might be dangerous to aircraft.

2.7 CASH TRANSACTIONS

A superintendent must have some cash or a checking account on the job. If the job is small, minor expenses may be paid out of your own pocket to be repaid later by your employer. If the job is in the same city as the office, layoff checks can be ordered by telephone on a few minutes' notice, and you need no checking account.

On larger or isolated jobs, the superintendent has both a cash and bank account. *Petty cash* is a small amount of money given to you when the job starts, and you request more money by sending the office a list of the payments you have made from the initial amount. You should have as much money at the end of the job as you started with. You may add to the petty cash, assuring yourself that there will be no shortage, by putting occasional job receipts into petty cash. These receipts come from cash payments for backcharges to subcontractors, sales of leftover materials, and similar sales.

Payroll checks may be prepared at the home office for jobs anywhere in the country if the payroll list is telephoned from the job to the home office. Payoff checks must usually be made at time of layoff. You should not have to keep unneeded workers on the job while waiting for payoff checks to arrive. For this reason, you usually have a bank account, separate from any account handled by the home office. Your signature must be validated by a corporate certificate given the bank by the home office unless the account is opened in your own name. Although only a few contractors use the superintendent's name on bank accounts, this is actually the safer procedure for the contractor. If the superintendent is authorized to draw checks on a corporate account, there is always the possibility that he may obtain a check made out to the firm and deposit it to the corporate account, whereupon he can withdraw it as cash.

If payoff checks are made from a bank account, checks are received from the home office with the next payroll. Although made out to the workers, they may be deposited by the superintendent to the company account. It is not commonly understood that a bank will accept for deposit practically any endorsement if it knows the depositor is good for the check if it fails to clear. If, for example, I find on the street a check made out to John Smith, I can endorse it John Smith and my own name and deposit it. Unless the writer of the check is not satisfied with the endorsement, there is no further question about it. If a superintendent has an account in his own name and it is convenient that paychecks made out to others be deposited in it, the bank may want a specific authority from the home office to accept such checks.

Although firms normally use a stamped endorsement for checks they deposit, this is not necessary. Banks will accept checks for deposit if they are made out to the person or firm who has the account. If you receive dividends or rent, you may have the checks mailed to your bank and account number, and the bank will accept them for deposit without any order from you. Also, anyone may deposit checks he has written to his own account. Usually, of course, there is no reason to do so except for record purposes.

Some banks will write paychecks as a regular service for a nominal charge. The superintendent may then turn in the payroll in rough form to the bank, who will compute the amounts and write checks. Cash is rarely used for payroll payment because of the danger of theft, and although labor unions often retain the right to demand cash, they seldom claim this right.

Company blank checks must be carefully stored, preferably in a safe. Blank checks, if stolen and filled in by the thief, are technically forgeries and the contractor has no obligation to honor them. As a matter of practice, if a contractor failed to honor such a forgery, merchants who cashed checks for workers would be more careful, making the company checks harder to cash. The workers, finding their checks unacceptable, would demand payment in cash, and the added expense and inconvenience to the contractor could be greater than the cost of the forgery.

2.8 TRUCK DELIVERIES

The superintendent should check with local delivery trucks and payment methods and notify the home office which lines offer best service. Some truck lines will deliver nearly anywhere, and others are limited to metropolitan areas. Some lines will allow collect items to be paid for monthly if arrangements are made in advance. This is a big saving in time, as delivery and payment in cash may cost more in time on the job than the amount of the charges paid. On a large job, a window is provided in the job office toward an incoming driveway, with a sign "Delivery trucks check in here" to expedite acceptance of shipments. If contractors' trucks must leave the job to pick up shipments, the Teamsters may claim jurisdiction over the drivers. Teamsters may have working rules that allow them to do any kind of work, directly conflicting with the jurisdiction of many building trades.

2.9 TYPES OF COMPANY ORGANIZATION

In this book, "firm" refers to a business, whether owned by an individual, partners, or corporate shareholders. The construction industry is nearly unique among industries in this country in that many firms are not incorporated–that is, they are operated directly by the owner or owners. If a firm is large enough to have a permanent office, forming a corporation has financial, tax, and insurance advantages. "Company" is therefore also used here when referring to a firm.

The supervision and opportunity for advancement in a construction company depends largely on the kind of organization. Generally, emphasis may be on *centralization* of authority–running everything from one head–or on *decentralization,* delegating authority as far as possible.

A *centralized* organization will often have an estimator who prepares estimates only, a purchasing agent who handles all the buying, and a field superintendent in charge of the work force. There may also be an office engineer for inspection and sometimes scheduling. Such an organization utilizes every person at what he can do best, so work is done as quickly as possible by the lowest paid personnel, and it is often difficult to locate the source of an error or the responsibility for a decision. Although large organizations are often centralized, this type of organization has not generally been successful for firms with a large number of medium-size or small jobs.

In a *decentralized* organization, the project manager's duties will include estimating, purchasing, and job supervision. Sometimes, the project manager has estimating and purchasing duties, but job supervision continues to be the responsibility of a field superintendent. If the job is a large one, so that the project manager can have but one job, he may be in the job office. With this method, the more highly qualified (and paid) manager has to do a certain amount of routine work, but responsibility is definite. The superintendent knows whom to ask to find out what materials are ordered or what is scheduled.

2.10 EXISTING STRUCTURES NEAR WORK

If your excavation for foundations or utilities is near another structure, you may have to build supports for the nearby structure. What is "nearby" may have to be determined by an engineer, but if the depth of your excavation is greater than one-third of the distance to another structure, you should confirm your intentions with the contractor. If the ground is to be pumped or well-pointed, damage to a neighbor may occur as far as 20 or 30 times your excavation depth from the excavation. If the excavation comes near neighboring buildings, support for the adjoining property may be required. Design of this support requires experience beyond the foreman level, and you should discuss your proposed method with the contractor. If he gives you full responsibility, you should:

1. Discuss the matter with the municipal building inspector and ask him to suggest a consulting engineer.
2. Ask the supervising engineer or the architect's representative for suggestions to be followed.
3. Engage a consulting engineer *experienced in this kind of work*—not just any engineer.
4. Check with your insurer to be sure your operations are covered, since you may be working outside the boundaries of your job site. Most standard policies exclude certain risks, and a special rider and additional premium may be required.
5. Obtain the permission of the adjoining property owner and approval of the engineer's plans or instructions by all parties mentioned above.
6. If at all possible, hire or borrow a foreman experienced in this work.

2.11 CHECKING PLANS

If you have considerable experience in the field, you will probably dislike working alone with a stack of drawings, looking for errors which you doubt are there and which you will eventually find anyway. But checking is one of the most important parts of the superintendent's work, and if properly done early in the job, there will be much less work, confusion, and expense later.

When you start a new job, take all available papers—plans, specifications, and material orders—and go over them carefully for as long as you can stand it. Few people have the patience to study a plan, or to visualize a structure, for days at a time. An estimator who does this has something to do other than just look at the plans. The estimator will often know of discrepancies in the plans, and you should talk to him first; estimators rarely complain about more than a small number of the errors they find, as they work so close to deadlines.

All dimensions have to be checked, of course, but more important, the details must be checked against the large-scale floor plans, and the various trades must be checked against each other. Errors will be there—they always are! Some errors are caused by changes in the drawings not being carried through all sheets, some because the draftsman leaves out dimensions on fabricated items because he does not know them, and many because the building is hard to visualize and relationships among various parts are not evident. No architect completes all details, and you should decide how you are going to build the items not shown.

As a foreman, you will have encountered many discrepancies on plans, but usually only those involving your own trade. Now you must consider all trades, and there are many more opportunities for error. Items that can cause difficulty are far too numerous to list, but some examples are:

1. Holes in structural slab are too small for ventilation ducts or are not located properly for partitions. This may occur because the thickness of insulation was neglected, or because nominal dimensions of partitions were used where actual dimensions were required.
2. Plumbing fixtures are shown on floor plans with no connecting piping on the piping drawing.
3. Inadequate space is provided in the ceiling for necessary piping and ductwork, and interferences occur between different trades, such as for piping or structural bracing passing in front of doors.
4. Inadequate space in partitions for specified door frames.
5. Details are shown for partition butting into windows at only one height above the floor; details at other heights are not shown.
6. Typical details are shown for running trim or overhang but without details at the corners where two types of structures run together.
7. Details or elevations are missing for portions of exterior walls that do not show on building elevations.
8. Stairwells and other multifloor facilities are built-in sources of plan mistakes from one sheet to the next and one trade to another.

If discrepancies in the plans are discovered early in the job, they can be corrected before they come up in the actual work. The superintendent with the answers is the person who asked the questions first, and the superintendent who works from poor plans will be blamed for a draftsman's omissions. If you are given sketchy plans, try to get them completed or redraw them yourself. Sometimes a contractor must hire an architect to complete plans that were good enough for estimating but not for construction. The estimator can and will work from *any* plans, by allowing for what isn't shown; but you must know (or decide) what is to be built.

You must not only find errors but must follow through by correcting your own prints and by notifying the architect and contractor, so that new drawings will be correct and purchases properly made. And *then* you must check all new drawings that you receive to be sure they are correct, and new purchase orders or material deliveries to make sure that the corrections have been made.

Don't hesitate to ask questions or report apparent errors because you aren't sure of yourself. If you're wrong, the contractor knows you're being careful, and it makes the estimator or architect feel good to think they're smarter than you are. Never be too smart! That can only get you in trouble.

2.12 PROFIT SHARING

Many superintendents are paid a percentage of the profit on their jobs—10 percent is a popular figure. When starting work for a new employer, do not expect too much from this arrangement, or take a job with a low salary, expecting to make it up on a profit share of what looks like a good job. There are many ways to figure cost, and unless a very detailed contract is made, the employer is virtually free to choose his own method of accounting. This is particularly true because company profits and therefore income taxes are sometimes made smaller by charging as much current expense to completed jobs as possible.

In addition to such recognized direct expenses as labor and materials, there are many indirect costs that may be charged to the job in different ways. Some are:

1. Charges for company-owned equipment.
2. Material transferred from other jobs.
3. Added labor cost for workers who are being paid more than is justified for the work they do, in order to keep them for future jobs.
4. Workmen's compensation, public liability, and other insurance.
5. Home office salaries and rent.

6. Telephone bills for calls from home office to job.
7. Travel expense of home office personnel.
8. Travel expense and salaries of supervisory personnel who visit several jobs.
9. Shipping costs for transfers from other jobs.
10. Salaries of owners of the firm.

It is usually unwise for an employee, especially one starting a new kind of job, to insist that the employer give a detailed list of these items. Consequently, profit-sharing plans have their greatest value for jobs after the first one, when custom has established what costs will be charged to the job. The job costs, for figuring the superintendent's bonus, may be different from the job costs used for income tax purposes. Keeping books one way for the employer and another way for income tax purposes is not illegal; it is illegal only if the books for income tax purposes do not show the information required by the government. Many employers attempt to keep their books confidential and still promise the superintendent a percentage. If you do not have full access to all charges involved, the bonus promise is no different from a promise of a raise "if things go well."

2.13 USING AN ASSISTANT OR HELPING A SUPERIOR

You have to convince the contractor that you are properly directing the personnel under you. This may lead to a situation in which you feel you have two jobs—you are one person to your subordinates and another to your boss.

In large organizations, it is not uncommon for all supervisors of importance to be duplicated—one person to do the work, and another to convince the boss or home office that the job is being done properly. This situation is likely to arise if the contractor or field superintendent relies on written reports, which he expects the job superintendent to prepare.

In smaller firms, the contractor relies on personal inspection of the job to see how well the superintendent is doing. Some contractors want to tour the job with the superintendent, to see how thoroughly he knows his job and to watch things being done.

This procedure keeps the superintendent from his work, and other contractors feel they will see only what the superintendent intends them to see, unless they tour the job alone. On an out-of-town job, the contractor may not be known to the workers and he may find out what is happening—not only how the work is going, but he can overhear remarks and discover dishonesty—and the contractor may learn more than you know yourself.

Don't resent the news he brings you. Although you are on the job constantly, you are not able to wander unknown and see what the workers do and say when they know you are not around.

2.14 THE HERO AND THE VILLAIN

It is a popular and effective method of management for the superintendent and contractor to take different views of transactions; that is, one is very agreeable and one is very nasty. This works so well that if there is no difference in their outlook, they pretend there is; one plays the hero and the other the villain.

For example, suppose that you have a boss who gives you full authority and always backs you completely—which he should do. If a subcontractor knows this to be the case, he has no appeal from your decision and knows it. If you make an error, or decide that some partiality should be necessary, it is very difficult for you to change your mind. You are the tough guy, and want to preserve that impression. But a subcontractor may appeal to your boss for an extra payment. Although he really is not entitled to it, you believe that if he gets it, he will do better on the job, and other subcontractors will feel less unhappy. So you tell your boss to overrule you and make the payment. The subcontractor is happy, and feels a little guilty for going over your head; he is likely to do better on the job to regain your friendship. Yet, you have not admitted to being wrong, as this would lead every other subcontractor on the job to cry on your shoulder; since *you* didn't make the payment, they won't bother you, and the boss is not so readily accessible. Besides, they know that their chance of going over your head is still slim—and they are afraid it will make their job relations worse.

Furthermore, if the subcontractor believes that he has some chance of appeal over your head, it will be easier to get him to sign orders for extras—although the orders may not state they are extra charges, it is much easier to charge for them later. Although you are forced to be a tough guy, the subcontractor, business agents, and suppliers feel they have another try in case something doesn't work out. You can threaten cancellation of orders for nondelivery, or threaten to cancel subcontracts, even though you know that such cancellation would result in slower delivery in the long run. If such threats work, you have accomplished your purpose; if they don't you can arrange for your boss to overrule you: the supplier can never be sure and will extend his best efforts to make delivery. If you made such threats and failed to carry them out, you would soon be ignored. If the boss overrules you, no one ever discovers you were bluffing in the first place. You wouldn't bluff at poker if you had to show your hand in a successful bluff. You are covering up your bluff in the same way by hiding your bluff's failure as a ruling of the contractor.

This same technique is often used in buying. The superintendent keeps an offer open by approving a purchase, but the contractor can then do a little bargaining on his own. The subcontractor, of course, realizes this situation and will attempt to make a final offer only to the person who has final approval of the purchase. A superintendent may be unable to obtain a true final bid price if it is thought that he does not have the responsibility to purchase.

2.15 BUSINESS FRIENDS

It is sometimes said that there are no friends in business. This is not so; only friends can *do* business. But the standards for business friendship are entirely different from the standards for hunting and fishing pals. In business, you deal with your friends; your friend is the person who makes it most profitable for you. Perhaps another word can be substituded–*contacts* is sometimes used. But as a superintendent, you are the person who must be a friend to everyone–to the business agent you fight with, to the foreman you fire, and to the supplier whose material you refuse to accept. While a superintendent must be stubborn–and sometimes rigid–the people with whom he deals must understand his reasons and know what they can expect of him.

Your first contact with a subcontractor may be to tell him, "If you do this, we'll be friends." In keeping these contacts you must sometimes be partial to someone. For example, you have a large job with two electrical contractors working, and have a number of electrical contracts which are bid by both electricians. You find that one electrician has been bidding consistently low, and the other has missed contracts to the extent that he will have to cut back his labor force and possibly leave the job. So you give him a contract, even though his bid is a little higher–in this way you preserve competition, and eventually will get your work done cheaper. If you ask the high bidder to cut his bid, he will expect you to do the same later. But you don't tell the low bidder.

So what happens? If there is a bid depository, the low bidder may have found out he is low and will suspect the other bidder of cutting his price unethically. It is very unlikely that the low bidder will suspect that you have deliberately paid more than necessary for a job. *If* the electrical contractors are in *collusion,* the bidder who doesn't get the job figures he's been double-crossed and will probably refuse to bid further in collusion. In any case, you are preserving competition and maintaining friendly relations with both people.

There are very few secrets in the construction business, and where possible you should be frank and truthful. If you aren't going to give out information, don't pretend to do so, delay, or give false information. You don't have to say anything, but if you do, it should be the truth. In some respects, an office worker can't afford the truth, but you can't afford anything else; any misleading statement is immediately revealed, since you are dealing in physical work. It is pointless to tell a salesperson you will think about buying his product when a moment later he spots a competitor's truck coming onto the job site. It is so much easier to say that you wish he had come sooner, or to sympathize that his product is of such high quality that you can't afford it; in both cases you keep your contact alive.

The common expression that a person "gave the business to a friend" implies a one-way friendship. Business friendships prosper because they are of mutual advantage; when a friend doesn't help your job and cut your costs, he passes to the category of a hunting and fishing pal, a drinking companion—but you don't do business with him. Some people don't mix personal and business friendships—a personal friend once hired is too difficult to fire. A good rule to follow is to avoid doing business with personal friends unless you and they are prepared to quit doing business and yet remain personal friends.

2.16 INFORMATION SOURCES

If you are reasonably friendly, you should develop your own sources of information about what affects your job. Did your supplier really order your material, or did he forget it? The owner of the firm may cover up, but the estimator or salesperson can tell you. Was one of your foremen verbally attacked in a union meeting last night, and with what result? Is a foreman treating blacks improperly? Is one worker "carrying" another? Is a competitor stealing your workers, and what does he offer? In your own firm, did a bookkeeper report something to the contractor—to your benefit or to your harm?

Making contacts to gather such information does not mean that you neglect your duty to your employer. There is usually no difficulty in getting workers to talk; the hard part is to make them stop. First, you must keep your own mouth shut; you can act on your information, but you don't reveal where you got it. Usually, telling what you know is also giving away your source—losing not only *that* source but others as well. Even your superior should not know your sources, as he may not understand what information can be repeated to others.

Where your own personnel are involved, your information (which is usually no more reliable than gossip) can rarely be used except to tell you what to watch. If it is reported to you, for example, that one of your foremen has taken home a power saw, give him an opportunity to report this fact to you himself without his being embarrassed. Perhaps he meant to bring it back, perhaps not. But until he has directly lied about the matter, he has not been dishonest. It may be that you would be happy to let him keep the saw rather than lose him, but this is no solution. If he has done something wrong, he will feel guilty and resent you—and quite possibly go even further past the bounds of honesty.

Also, this information gathering must not be allowed to break up your crew. If a worker offers information about a noncriminal action, he should not be given any special consideration and if you use the information, he should never know it. It is wise to indicate that his report was similar to others which had reached you—this conceals the source of your information, even from the informer. This information may or may not be malicious—this is immaterial. If a labor foreman reports that another foreman is improperly preparing the ground under foundations, for example, the report is the important thing; it may be that the foremen are enemies, or they may be friends. You check the report at the earliest opportunity by actual examination, and can take appropriate action based on your personal knowledge. If you get reports that a foreman has an antipathy toward blacks or other groups, you would try to keep him from jobs where he would be required to deal with these persons. If a bookkeeper has reported that your job is unprofitable and you believe the report to be in error, you can ask for a report or prepare your own for the contractor, without mentioning that you know a report has already been made.

2.17 THE INDIVIDUAL CONSTRUCTION WORKER

Professional management consultants and college professors have written a lot of pamphlets, books, and magazine articles about how workers should be treated and how they feel about their superintendent. Most of these publications refer to unskilled or semiskilled people, or to skilled workers who don't have an option to quit. It is hard for an electronics expert or a railroad trainman to find another job, and his attitude is affected by the fact that he knows he can't walk out. The construction craftsman has no such restrictions. If he doesn't like you, he can find just as good a job across the street. You are in constant competition for workers; even when there are a number of workers on the street, you are still looking for the good ones who don't remain out of work very long.

There are many other ways in which the construction worker is different from other workers. His job cannot be readily automated out of existence, and it is hard to split a skilled job up among a number of unskilled people. The best *craftsman* (except in the trowel trades) has many skills, many of which can be learned in a few minutes. But it would take many more workers than could be used if the work was broken up, and there would be no continuous work for any of them. In terms of the number of personnel working at one spot, fewer people work together in construction than in any other large industry.

This means that supervision of a construction job is a personal thing; each worker is well known to his supervisor, and managers off the job are nearly helpless in attempting to supervise the work. "Management experts" are unable to change this situation and often react with bitter criticism because construction operations are "old-fashioned, obsolete, and unchanging." They criticize unions, workers, architects, bankers, and building inspectors as a group, usually to claim that if the stubbornness of the people concerned could be changed, new and better materials could be used.

Unfortunately, most of these new and better materials are more expensive. Those which are cheaper are used readily enough. It is all but forgotten that the use of concrete frame buildings was at one time bitterly opposed by the bricklayers, to the point that some locals refused to work on masonry curtain walls on concrete frame buildings. This obviously did not stop the introduction of concrete buildings. Like concrete, the new materials will be adopted when they are cheaper. The older materials—concrete and masonry products—cost as little as 2 cents per pound in the completed structure. These products are dug from the ground and run through comparatively simple and completely mechanized operations—it is not surprising that substitutes are hard to find.

Many hand operations have been developed over a period of centuries. Frank Gilbreth, an early pioneer in time-study methods, began by studying existing methods used by bricklayers. As a general contractor he obtained high labor efficiency chiefly by adapting the best existing practice to all the workers he employed; in later years he became a consultant in manufacturing operations, applying what he had learned from bricklayers. In short, construction operations change very little, because they *are* old; they have become standardized over a very long period. The newer industries are still changing their methods because they have had less time to figure out the cheapest method.

There are many operations that can be improved, but many superintendents are not using the best methods available because they don't know about them. Labor cost cutting is accomplished in two ways: by improving your own method, and by copying any better way you can find. Usually, you find that the way you thought of has been used by someone else for years, or discarded long ago. Good mortar, for example, is what the bricklayer *says* is good mortar; don't try to be too scientific about it.

Construction workers are different from others in yet another way:

they are sometimes workers, sometimes foremen, sometimes contractors. It may be hard to get a foreman to really tell his workers what to do because the foreman expects that on the next job the worker will be *his* foreman. Craftsmen usually know what the foreman is doing and know how much production is expected. If you do not demand a day's work, and if you are not careful of your employer's material and equipment, you lose the respect of the workers. If you do not properly schedule contractors (as, for example, calling the lather to the job when insulation behind the lath has not been ordered), all the workers on the job, not just the subcontractor involved, know that you goofed. Don't try to to cover up your mistakes—workers know that everyone makes mistakes and you need their support. If they think they must work harder because you're nice but not very bright, don't argue the matter—the important thing is that they're trying. If you have personnel working for you who are older and more experienced than you are, don't make a contest between you. You may find that if their experience is recognized, they are quite willing to help you all they can. It is often a mystery to a young person on the way up that an older worker, working at a job below his past experience, is perfectly happy with it. This is particularly true if the "better" job actually pays in prestige but has a lower hourly wage. Don't try to make others into your own image, but take advantage of each person's virtues and try to put him in a job where his faults are not important.

2.18 ETHICS

Any employee is expected to be *honest* and *good;* otherwise, he'd be a criminal and not working at all. The more exact definition of what is *honest* is *ethics.* Ethics defines what one can do in business as it *appears* to *others.* If you *appear* to be dishonest or someone *thinks* you are unfair, the effect on your employer and ultimately on yourself is the same.

Although there are many ways a superintendent can embezzle money, practically all such operations involve more than one person. A superintendent can obtain payments without collusion with others only if the management of the firm is loose—if the contractor does not check the purchases to see if anything has been bought twice. If a superintendent is handling bids from subcontractors and preparing purchase orders or subcontracts, he can recommend the purchase of the same item twice—under different descriptions—and recommend one of the purchases to himself under another name. He then acknowledges receipt of the material, or approves completion of the work in place, and bills the company under a fake letterhead. Since many small firms do not even have a letterhead, even this amount of preparation may not be necessary. This requires that the contractor be ignorant of both the work and the subcontractors and suppliers with whom he is doing business—which does happen.

But this is a criminal act against which there are adequate laws. There are other things you may do which are not illegal (although some of the examples depend on laws in your state) but which give you an "unfair" advantage or which divert some of the employer's money to your pocket. Since ethics is a matter of judgment by the parties involved, no one in the construction business can say that he is entirely ethical, as judged by those outside the business.

In the professions, ethics are supposed to protect outsiders, but as a practical matter they are most often applied to protect other members of the group. In many of the learned professions, advertising is unethical, and competitive bidding is a serious offense! It is unreasonable to say that "everybody" in the business is unethical—since ethics is just what "everybody" says it is!

The situations you are concerned with depend on *who is being hurt.* The most common situation is the acceptance of unsolicited gifts from subcontractors and suppliers. These gifts are part of their operating expenses and ultimately are paid for by their higher prices to your employer: theoretically, your refusal of gifts will ultimately result in lower prices. But this theory is idealistic; actually, the usual ham, bottle of liquor, or similar gift will be charged to all employers alike, and it is unrealistic to assume that a saving will really return to your employer. This kind of "commercial Christmas" can be diverted to more productive purposes only by common action—as in Jacksonville, Florida, where local subcontractors agreed to contribute to a scholarship fund instead of giving presents to architects and general contractors.

But suppose you solicit subcontractors for contributions to a favorite charity of yours? They are contributing to a cause only because they may not give you the money directly. And where does the acceptance of Christmas gifts end, when you are obviously getting more than a firm gives to all people in your position? The U.S. Army Corps of Engineers once specified that their engineers could accept only gifts with the name of the donor firm printed on it—that is, advertising gifts such as pens, ashtrays, and calendars. One wag requested that the giver's name be painted on his Cadillac with a not-too-permanent paint so that it could be removed easily. Another rule requires that gifts be distributed among all persons in an office so that no one can profit from having shown favoritism to a contractor.

Since you are not working for a public agency, your problem is much simpler—as long as your employer is informed of what you are receiving, he may decide if it is proper, and usually no one else is involved. You may do things that would not be proper at all if working for a public agency—such as accepting part-time employment for firms with which you are dealing in your regular job. If you are working for a large firm, it is necessary to have stricter rules; the contractor may not feel that *you* would be affected by certain gifts or payments, but that other superintendents would automatically do the same thing you did.

A much more difficult situation arises when you must decide if you are going to follow your employer's policy when you think it is unethical. You can become identified with a particular employer's policy to the extent that you will have difficulty in either getting a job in the locality, or in engaging in construction work yourself. This is a personal situation, and often the best solution is to resign. No listener will admire you for talking about your employer to him even though he agrees with you, so you should not discuss the situation with people not employed in the firm. You may choose not to engage in the portion of the business where the contractor is not ethical, particularly where bid-shopping is involved, and to make it clear to subcontractors that you only build the job, you don't buy it. You may make it clear that the orders you give to subcontractors originate in the home office; you need not and should not criticize the orders, since the subcontractors will do this well enough themselves.

You must also realize that what may appear to you to be unethical may be local practice in another area. In one area, subcontractors may customarily give their *best price* (actual lowest price) at bidding and refuse to cut their price after contract award; in another, the first bids received may be quite openly just something to talk about, and no one takes them seriously. If your employer is well acquainted with unethical practices of an owner, your employer may be engaging in some hanky-panky of his own which he would not do if an ethical owner were involved. Since an owner expects to deal with construction firms only once, in many instances he is much more likely to act unethically than is a contractor, who has a reputation to maintain. Vendors who bid custom work or lump-sum material prices may also lower bids later, or bid only after contract award.

Out-and-out commercial bribery—an offer to pay you to reveal bids or to recommend a particular bidder—can often be avoided by a refusal to recognize an offer of bribery. That is, *act stupid.* If a subcontractor wants to know what he has to do to get a contract, just tell him to lower his price. If he mentions a figure he would be willing to pay, interpret this as a deduction from his bid. A calculating attitude often attracts such offers when an open attitude does not. In one instance where two people were dealing with subcontractors in the same city, one regularly received offers of bribes while the other never did.

Rarely, you may find an inspector frankly on the take. One Corps of Engineers inspector proposed to a contractor to make a certain payment to assure that the monthly payment was adequate. The project manager refused and left his own estimate with a note that he would like to talk to the resident engineer about the estimate if the payment approved was any less than the project manager requested. The payment went through as requested, and no action was taken. The contractor in this instance refused to report the inspector (after all, there was no documentary evidence) but did instruct all his own employees that they were never to be alone with this inspector—even telephone calls from him were monitored by the switchboard operator. Eventually he was caught in another matter, as expected; the contractor did not want to be involved with him in any way. This inspector went so far as to offer unsolicited cost information on "confidential" military jobs to prospective buyers.

Dishonest inspectors should be reported, but how? Since the inspector's superior may be involved, it is well to go as far up the line as possible, with a question such as: "Are you sure of the honesty of your inspectors? There seem to be some stories running around." If the supervising engineer launches into a defense of his inspectors, it is well to drop the matter. The engineer can get straight answers if he wants them by asking subordinates and contractors—many people will not report such matters but will not lie if asked directly.

Are you obliged to continue in your present job when you have a better offer? You usually have no legal obligation to give notice. However, few contractors can readily replace a superintendent in the middle of a job, and superintendents consider themselves committed to complete the job for which they are hired. They must do this in order to get a job later, either for the same contractor or for others in the same area. Also, it is essential to have a satisfactory statement from past employers. But if your employer threatens to fire you in the middle of a job, he shows that he does not feel obligated to keep you and there is no reason why you should feel obligated to work for him.

There is also no reason why you should not be looking for a job while employed; in *fact this is the best time you can look for work.* Superintendents' jobs don't start just anytime, and there may be only two or three openings a year, even in a large city. If other employers want to talk with your present employer about you, they can do this without revealing that you have been actively seeking work—they just inquire if you are going to be free after your present job. This doesn't hurt you a bit, and it emphasizes to your employer that others are interested in you.

Sometimes a contractor dislikes having his superintendent disclose his salary to others. You have nothing to lose by advertising your salary; if it is low, you are encouraging job offers; if it is high, you are demonstrating your value. Of course, you may agree with your employer and be as ashamed of your earnings as he is! Besides, there are few secrets on a construction job—or in the construction business.

Some employers feel that a superintendent should not talk about a previous job with a new employer. It is wise to give information only when it will do you some good, and not to distribute it so it will be repeated to a former employer. You are not obligated, however, to keep quiet about anything after you have changed jobs; few people do.

In short, good ethics consists of acting toward others as they expect you to act, and acting so that they will continue to do business with you. If you can improve on the system without increasing your job costs, by all means do so; the superintendent is usually in a much better position to be frank and honest than are people who have more information.

Contractors often quote owners a list price for items, particularly electrical fixtures, rather than cost price. This is not usually illegal if done by subcontractors, since the owner has the option of purchasing such items elsewhere. The standard contracts prevent the general contractor from adding a markup on these items, but do not mention the subcontractor, and special list-price catalogs are printed for the purpose of submission to the architect for selection of some of these items. There should be no reason for you to be involved in this practice, as the subcontractor usually deals with the owner directly. It should be eliminated by specifying minimum alternative acceptable items in the first place and negotiating changes for items that are more expensive than specified.

2.19 COMPLETION OF JOBS

Starting a job is fun—spending money (especially other people's), making new acquaintances, finding new information, hiring new people, interviewing job applicants, building an office, growing. Finishing a job is a bore—bickering with the architects and subcontractors over completion of interminable details, firing your friends, making excuses to the boss as to why you forgot this or that item in your estimated Cost to Complete form.

But this is an important time to the new superintendent—are you staying with the same contractor? You may have done a good job until this time, and then get cranky and fail to follow up details as you should. You are quite likely to get discouraged toward the end of the job, because you don't think you are earning your pay—and in this business, a person who thinks he is worth less than he's getting often quits too soon. But your boss has been through this many times—he knows there is an unavoidable waste of time at the end of a job. He wants you to annoy the subcontractors to get them to finish details so *he* won't have to do it.

Many superintendents, especially journeymen who have not had much supervisory experience, try to keep their new office clothes clean—to remain supervisors with nothing to supervise. Foremen have a habit of hanging on at the end of a job when they are no longer needed—sometimes because of union rules. They, too, don't want to swing a hammer after telling others what to do for 6 months. So one day the contractor drives up to your job and sees a superintendent, doing nothing, supervising a nonworking foreman who has but two journeymen to supervise. Since the contractor is there himself because *he* has nothing to do, he may decide to fire everyone in sight to shake things down to normal.

Some contractors will never demote a worker, either in position or in pay, as they believe the worker will think he should have a better job and will not produce. So if you're out of supervising to do, you may have to offer to work with your tools (or with your pencil, if you're an estimator) to avoid being automatically laid off. A wage reduction is another matter—your wage is originally set with the idea that it will be either temporary or permanent. If you take a wage cut when working with your tools, you can expect more money when working as a superintendent than you would otherwise get. Most contractors keep superintendents on a straight, fairly low wage which is constant for all kinds of jobs they may hold but which may be supplemented by a bonus at the end of each profitable job.

3 Material and Subcontractor Payments

3.1 THE BOOKKEEPER

The bookkeeper of your firm is a very important person. He watches over your employer's pocketbook, and regardless of its apparent physical location, the wallet of a business owner is quite close to his heart; otherwise, he doesn't remain in business. While it may appear to you that the bookkeeper worries over some quite unimportant details, remember that when the bookkeeper worries, the boss worries. And the boss will not be doing as he should—worrying over the size of your Christmas bonus!

So get together with the bookkeeper and find out just what you should do for him. And when his call from an air-conditioned office gets you out of a hot, muddy hole to explain what happened to a concrete ticket from 2 weeks ago, don't say what you're thinking. You *want* him to call you, without mentioning the matter to his superior. The bookkeepers can do very little *for* you, but their complaints of irregularities can do a great deal *against* you.

If necessary, go to the office periodically and see what the bookeeper does with the papers you send him. Don't appear cross or abrupt—he's like anyone else, and wants to talk to people who appear to like him. If he thinks you don't like him, he'll talk to someone else—to your detriment.

3.2 MATERIAL PURCHASE PROCEDURE

Many contractors, not necessarily small ones, are very sloppy about material purchases. The worst procedure is that in which a manager orders material by telephone, the superintendent signs a delivery slip, and the bookeeper pays the bills—with no verification of the invoices among the three persons. Do what you can to avoid being part of such a procedure, as sooner or later mistakes will occur and some of them will be discovered. The contractor is then most likely to blame anyone but himself.

So keep a record of your own purchases, quantities, and prices, and send all delivery tickets and invoices to the office promptly—you will then be doing your share. The steps outlined below should be done, one way or another, for profit and tax purposes, and above all on *cost-plus* or owner-paid work. To understand what the bookkeeper does, you must understand that a number of decisions are made in regard to each material item before it is finally paid for:

1. The quantity and description (specification) is needed. This is done by the estimator before bid, but this estimate is usually checked and made more detailed.
2. The price must be determined—by a purchasing agent or project manager, or often by the estimator, who takes bids and writes the purchase order. This purchase order is the written order to the firm supplying the material.
3. It must be delivered—someone has to decide when it is needed and to see that it gets there. This may be done by the estimator, superintendent, or both together.
4. It must be checked for quantity and quality—this is done on the job, by the superintendent or clerk.
5. The payment must be made—the billing and quantity are checked against the original order and quantity received by the bookkeeping department.

All these actions are required even if you pick up a phone and call for a two-by-four. They are not papers—they are decisions. If the price is not quoted before delivery, as is frequently the case on small orders, the person checking the invoice for payment must decide if the price is acceptable.

Various papers are used to assure the bookkeeper checking the payment that all the necessary authorizations are completed. This does not mean that the bookkeeper considers himself smart—on the contrary, the bookkeeper who writes a check realizes that he knows nothing at all about whether the material was needed, whether it was delivered, or whether the price was right. He therefore wants the persons designated to make these decisions to tell him everything is in order. He can then be sure that no one will blame him for paying out money when he shouldn't. Therefore, he prefers everything to be in writing. And if it should turn out that fraud is involved somewhere along the way, only a written signature is useful for proof.

3.3 THE PURCHASE ORDER

A purchase order is a written order similar to a mail order blank. It is sent to a supplier with instructions on when to ship and where to deliver. The purpose is to order material, to make the price and quantity definite, and to specify other terms of purchase, depending on the kind of material.

The original order is sent to the supplier and a copy is kept by the project manager or person making the purchase. In a large firm there are other copies; the superintendent may get a copy, and if the bookkeeper is to keep track of orders and to *check* invoices, he needs a copy as well.

The superintendent should be given a copy so he knows that the material has been ordered and whom he should call for delivery. Often the superintendent's copy has the prices deleted, to prevent the unit prices—which may be quoted for a particular job or contractor—from becoming known. A supply firm may keep its prices secret, even to the extent of billing a price more than the agreed price, so that even the people who check the invoice never know the true price. In such cases, the supplier is quoting a special price for a job and doesn't want to be embarrassed by one customer who finds out that he is paying more than another. There is usually a set contractor price in an area, but contractors get lower prices on large jobs, according to the demand for material at the time the job is quoted. Often suppliers agree on the prices to be charged, but they may also agree that orders over a certain amount are to be bid competitively. Usually, they do not quote identical bids on public works, especially federal jobs, as this may leave them open for prosecution on antitrust laws. Such agreements are illegal in most cases, but the smaller ones are seldom prosecuted, and thousands of cases of identical prices exist even on public bids.

These agreements are not written and are sometimes not specific. Salespeople will deny they have an agreement but are obviously familiar with competitive quotations; often they speak of "meeting the competition" or "charging the market price," which is an informal agreement. However, if you are purchasing, never accept the fact that such agreements exist; it is always possible that you will break the previously agreed price, particularly on large jobs.

In a legal sense, a *contract* is an agreement, written or spoken. Every material order is a contract, usually including an *offer* and an *acceptance.* A purchase order may be an acceptance if the seller has made an offer. But if the purchase order isn't an exact acceptance, it is another *offer* and the vendor is released from his proposal. The vendor may accept the purchase-order offer by delivery, a letter of acceptance, or by signing and returning an acceptance attached to the purchase order.

The purchase order may not be for a definite quantity but for material to be ordered from the job by the superintendent. Blocks and brick, for example, are often ordered this way. The purpose of the purchase order is still to establish the price and to notify the superintendent from whom to order delivery.

3.4 FIELD PURCHASE ORDERS AND REQUISITIONS

Many firms give the superintendent a pad of short purchase orders so that he may write, in pencil, orders for small items. This is done to discourage verbal orders by the superintendent. Sometimes *requisitions* are used, particularly by larger firms; these are forms sent by the superintendent to the home office, requesting the purchase of some item. A requisition may have the supplier and price or it may not; the purchasing agent may fill in his own supplier and price anyway. If the material is available from the contractor's own warehouse, the requisition, in this case called a *warehouse requisition,* is all that is necessary to get the material to the job. The warehouse requisition serves as an order to the bookkeeper to charge the job with the material.

3.5 RECEIVING MATERIALS

The purchase order received by the bookkeeper tells him how much is ordered, from whom, and at what price. He still doesn't know if the material was received on the job, or if it was satisfactory. For this he needs a note from the superintendent; usually the superintendent sends a signed copy of the delivery slip to the bookkeeper. This delivery slip comes with the material; the person who made the delivery keeps one copy and gives another to the superintendent. It is important that the delivery slip show *clearly* the quantity delivered and the material; if it is a duplicate, you may have to go over it with a pencil to make it legible. Also, your signature should be clear, and there should be enough information to enable the bookkeeper to identify the purchase order. On a large job, you may have to put purchase order numbers on all delivery slips so that the bookkeeper can identify them.

This delivery ticket is a *receiving report* for the bookeeper, as it tells him the material has been received. Some firms, in order to keep receiving reports uniform, use a special form of their own in addition to the delivery slip, and others use such a form only when a delivery slip is missing. When a shipment is delivered by freight, for example, the paper received may show only the number of boxes and the weight. If the count and exact material is not shown, the bookkeeper doesn't know just what has been received and needs a material list. This list may be packed in the boxes, in which case it is called a *packing list.* The packing list may be sent to the bookkeeper along with the freight receipt, with a notation of any missing material. A delivery slip and packing list differ only in that the superintendent is asked to sign a copy of the delivery slip for the supplier; he receives but one copy of the packing list, and does not return a signed copy.

3.6 CHECKING SHIPMENTS

A bookeeper would like all shipments to be checked exactly—that is, item by item as shown on the purchase order. This is possible on most items; brick and block, or lumber and doors, can be checked by the number of pieces, counting the number of pieces wide and multiplying by the number of pieces high. Even here, though, some approximations must be made. Some pieces of lumber may be short, and you count only the ends. There are many different types of doors, and you count only the number. If hardware comes in sealed crates, you need space to spread it out and repack it to be sure all the material is there; you may not have the time.

It is customary to sign for shipments by the number of crates, not the contents, and you may claim damage later upon opening the boxes. If the supplier delivers the material himself, he may give you a packing slip to sign—a delivery ticket listing the actual items. In this case, be sure you don't accept the count unless you have an opportunity to check it. Otherwise, sign *quantity subject to later count.* If the person making the delivery objects to this qualification, offer to help *him* unpack the goods and count them. If he delivers you "eight dozen hinges," you are entitled to see every hinge; if he delivers you "one box hinges," you need sign only for the box. Few firms are dishonest, but all have clerks who make mistakes in packing.

The same qualifications must be observed with the receiving report which you send your bookkeeper. A *delivery slip* is received with shipments, but this often lists only the number of boxes rather than the number and description of items. For this reason, many firms use a *receiving report.* This is a sheet of paper on which a complete quantity and description account is written, to send to the office along with the delivery slip. Any damage found is shown on the receiving slip, and if discovered on delivery, on the delivery slip as well. Don't copy from the packing bill packed in the box without making your own count. If you haven't time to check the whole shipment, send in the packing bill with a note that the box received appears to be the same as the packing bill calls for, but don't say you checked it if you didn't. The home office is aware that your time and should be checked just as is delivery of material, as it is an important part of the order which must be delivered in advance.

3.10 SHOP DRAWINGS

Fabricator's shop drawings show the exact dimensions and materials which are usually lacking in architect's drawings. The architect's approval of these drawings merely means that the architect does not see anything wrong with them—*the architect does not check dimensions.* He does not note possible conflicts between trades or between different shop drawings unless the conflicts are obvious. Since he approves drawings ". . . subject to specification requirements," the architect's approval really doesn't mean anything to the contractor. It is the duty of the contractor to submit shop drawings, but the architect has very little obligation to discover errors. Contractors often fail to check these drawings—and find discrepancies later. Many contractors do not even unfold the drawings in their office but send them right to the architect; errors must then be discovered much later by the job superintendent. The well-managed office gets a copy of shop drawings to the superintendent as soon as received.

The superintendent is not expected to check internal dimensions of shop drawings; for example, he would not be concerned with fastenings for stair construction (unless he had to erect the stairs) or with the kind of railings used. He should have a set of shop drawings for checking the points at which work of various suppliers fit into each other. He should check story heights and the details of the concrete work where it is to fit steel stairs, for example. If he has shop drawings for precast terrazzo treads and for steel stairs, he should be sure they will fit together. It is possible that the detailer of terrazzo stairs would not have the shop drawings for steel stairs. Sheet steel sills should check against details of steel windows. Toilet partitions should be checked against actual rough-in and partitions in place; or if the partition drawings are made first, the partitions and plumbing rough-in should be checked against partition drawings.

Many costly errors are the responsibility of the general contractor, by reason of his failure to furnish one supplier with the shop drawings of another. Air conditioning equipment is often shown as schematic (approximate size) on engineer's drawings, and the plumber may install floor drains that do not fit the actual location of equipment. The electrician may also rough in conduit which interferes with air conditioning equipment. The sprinkler fitter's drawings must be available to the plumber and steam fitting contractors for them to properly locate their work. The purpose of providing the general contractor with shop drawings is to avoid such errors, and the project manager or job superintendent should make

is expensive, and even if you have a clerk, it may be more expensive to check these shipments in detail than to accept occasional loss. Your firm will set the standards for inspection. It is important that you don't sign any document if you don't know it to be correct; your signature is not merely a notice that you've seen the paper—the office must rely on it as evidence of what has happened on the job.

In the illustrated receiving report (Fig. 1), the superintendent has received a shipment of machinery parts about which he knows nothing. He has therefore obtained approval by the mechanic that the parts are those expected, and has so noted. Normally, the bookkeeper will pay bills if it appears that the shipment is substantially correct; reputable firms correct any errors later, and in many cases will replace missing items even though they feel the items were lost on the job. Keeping the original case closed

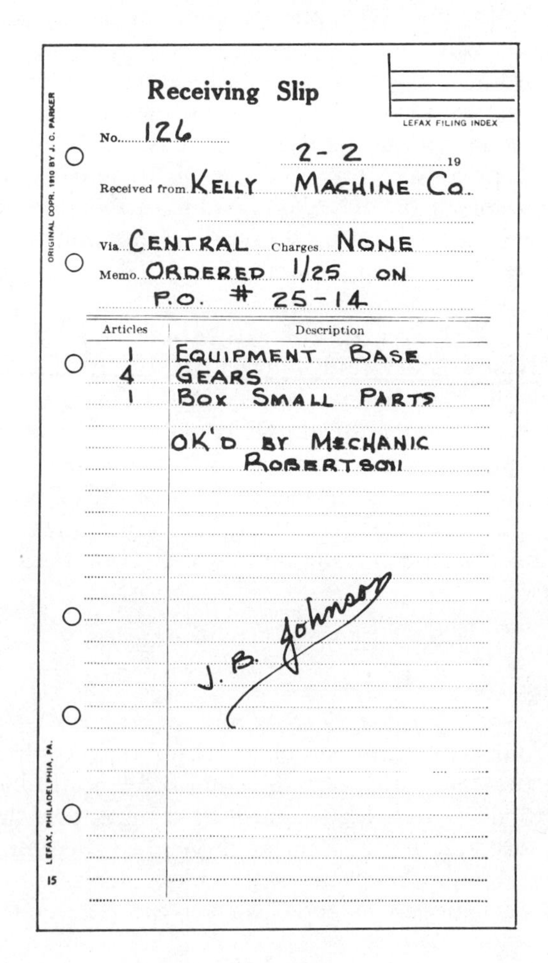

Receiving Slip

LEFAX FILING INDEX

No. 126

2-2 19

Received from KELLY MACHINE Co.

Via CENTRAL Charges NONE

Memo ORDERED 1/25 ON P.O. # 25-14

Articles	Description
1	EQUIPMENT BASE
4	GEARS
1	BOX SMALL PARTS
	OK'D BY MECHANIC ROBERTSON

J. B. Johnson

ORIGINAL COPR. 1910 BY J. C. PARKER

LEFAX, PHILADELPHIA, PA. 15

Figure 1 Receiving report.

and sealed until an actual count is made will both reduce pilferage and help locate the reason for shortage. Your firm will probably have standard methods for reporting shipments received.

It is important to remember that you should not order materials by telephone or in person without a confirming written order. A verbal order is an invitation to trouble; not only is memory unreliable (your own included), but a conversation, even if proven, may not constitute a contract. This depends on your local state law, and is of importance primarily if you are relying on a quotation as a basis for a bid of your own.

3.7 INVOICE APPROVALS

An *invoice* is a bill from the supplier to your employer, asking him to pay for materials. You will almost certainly be asked to approve invoices for materials you order, and you may be asked to check and approve other invoices. The practice of suppliers varies; some of them combine the invoice and the delivery ticket (the corner hardware store is likely to do this), showing all prices and the total on this ticket. Others mail invoices daily for the purchases of that day, sending a separate invoice for each delivery. Others mail invoices at the end of the month for all deliveries for the month. It is becoming common to mail invoices daily and to send no other bill at the end of the month.

The invoice, after you approve it, is checked by the bookkeeper against the purchase order and delivery ticket, and if any invoices are missing, he will call you. He may file the invoice in a separate file numerically, or under the name of the supplier; or he may also file it under the purchase order number. His filing makes a difference to you if you are asked by a supplier to locate his payment. If the invoices are filed numerically as they come in, it may be impossible to find out how much money altogether is due a particular supplier—which is of little interest to the contractor anyway.

3.8 STATEMENTS

A *statement* is an accounting summary from the supplier to your company, showing the amount due at the end of the month. Some firms send a statement with all invoices for the month, at the end of the month; some send invoices daily and a statement at the end of the month. Other firms do not send a statement at all unless the customer asks for it. Generally, statements are of little use except as a bookkeeping control, as they do

not show exactly what materials and purchase orders are covered and may include several jobs.

A statement does disclose the fact that an employee is charging material to the company and paying for it himself. This is the way many mechanics buy tools, for example; but it may also be the way a foreman is using his employer's credit for other work. If the foreman fails to pay, the contractor may be responsible. If you approve a statement for payment, it is quite possible that the material will be paid for twice, since all invoices, whether paid or not, may be repeated on the statement. For this reason, it is important that *only one copy* of delivery slips be submitted to the office; if there are two copies, one can be paid with the invoice and one paid with the statement. Statements are not always clearly marked as such.

The statement, since it lists the invoices, will give the bookkeeper a check on any missing invoices. If the bookkeeper has sent you an invoice for approval and cannot find it when the statement is received, he may call you for it. The statement is protection for the contractor that his workers have turned in all the invoices they have received; he has no other way to tell what may be outstanding. Firms can get into serious difficulties due to the existence of large outstanding invoices of which the accounting department was not informed.

3.9 MATERIAL EXPEDITING

To be sure materials are on the job as needed, the supplier is called before the material is due. This is *expediting*. On small jobs or small orders, the superintendent may be expected to check on material delivery himself; on larger orders, it may be done by the project manager at the home office. In any case, the supervisor on the job must know, and you should take over expediting as the delivery date approaches.

If you have to do all expediting yourself, keep a file of purchases in order of required delivery—that is, with the first materials needed on top. Periodically, go through them for a month or so in advance, calling those people who have special fabrication to do and thus require time to get the material ready. If the material is available from warehouse stock, it is necessary to call a week or so ahead to be sure the material is actually in stock and to remind the supplier that you need it.

Many fabricated items—such as miscellaneous metal, toilet partitions, sheet metal doors, and steel stairs—require shop drawings which are submitted to the architect for approval before fabrication is begun. Delivery date of these shop drawings should be shown on the purchase order

full use of shop drawings. Leaving minor details such as the temperature rating of sprinkler heads to subcontractors can cause considerable embarrassment later.

3.11 SPECIFIED DELIVERY DATES

It is often more important to avoid delivery of materials than to obtain it. Suppliers are very prone to ignore such instructions on purchase orders as, "Do not deliver before June 1," for materials which are ordered early and are not needed until the end of the job. Such early delivery takes up valuable storage space and may tie up working capital. If early materials are delivered, they should be accepted subject to payment at a *specified* delivery date rather than at the *actual* delivery date. If there is a shortage of space or of storage protection, refuse delivery. You will not be able to refuse delivery, however, unless the earliest delivery date is specified in the purchase order.

3.12 SUBCONTRACTOR PAYMENTS

Many types of construction, such as electrical, plumbing, and roofing, require highly trained workers. Most general contractors, who agree to complete the entire project, cannot keep these people working because they are not needed much of the time. The *general contractor* (often called simply the general) therefore agrees with firms who do only one kind of work to complete the work on each job. These firms agree to perform part of a building contract and are therefore termed *subcontractors* (often called simply subs).

Subcontracts are usually made at the home office, and the superintendent has little to say about them. If the superintendent handles the subcontracts himself, it is important that he specify the progress of work in the contract, by specifying the number of workers or the dates of completion of the work. Such matters as services to subcontractors (water, trash disposal, electricity) should be included, and material storage and cleanup should be clearly explained. It is very poor practice to use "one-time subcontractors"—that is, firms with which you will do business only once. These firms are often either accidental low bidders or are new firms with little money behind them. Acceptance of the low bids, without investigation, often results in a job that is greatly delayed by bankruptcy of subcontractors. Even if the subcontractor is bonded, the confusion of schedules and the delay in completion can cause both direct and intangible serious loss to the contractor.

Subcontractors are usually paid monthly, based on their own estimate of the work done. The general contractor, in turn, is paid by the owner for the work completed by the subcontractor and takes little interest in the amount the subcontractor requests. In fact, the more money paid to the subcontractor, the more the contractor may get in advances on the job. Subcontractors, like contractors, find it difficult to finish a losing job on which they have overdrawn, as in the last part of the job they are putting in more money than they are getting paid. Since they have the money in another job, they must complete the other job in order to get the money to complete yours. For this reason, it is well to give the subcontractor an incentive in the form of an overpayment near the end of the job, not at the beginning. Overpayment at the start of the job can result in slow completion.

Some general contractors use payment as a threat to subcontractors to get work done on time or to get work done without an extra claim. This is an unfair practice; the subcontractor in many cases is so short of money that he will agree to any terms in order to get his payment. However, he will not be anxious to do business again with that general contractor, and may charge more or refuse work entirely. Also, agreements made under such pressure may later be disputed in lawsuits; if a subcontractor agrees to a change in his contract without being paid for the change, he may later refuse to honor the change on the grounds that there was no payment—you merely promised in return to do what you already had to do.

The amount of a subcontractor's payment is greatly influenced by the number of personnel he has working and the status of his material, however. You really have very little to go on to determine the amount due the subcontractor. Since there is a lag—often as much as 10 days—between the approval of his bill and his receipt of payment, you can overpay him considerably at the time of approval and he will have caught up by the time he actually receives his payment. If he has but one or two workers on the job and is bringing material from his warehouse daily, you should be more careful about the size of his payment than you would be if he had 20 men and a yard full of material on the job.

Subcontractors may often be paid for the materials they have stored on the job; the amount of the payment probably depends on your particular contract with the owner. It is a good practice to encourage the subcontractor to get his material out on the job, as only then can you be sure he has it. If the subcontract is properly written, it is theft for him to remove material from your job to another after it has been paid for; sometimes he may not remove it even if it has not been paid for. On occasion, he may be paid for material that he has not delivered to the job; in this case you should pay only for special materials that cannot be used on other jobs and require covered storage or locked storage not available on the job. You should check such materials personally to be sure they are actually in storage.

If a subcontractor cannot collect for material delivered to the job, he will try to schedule his material delivery so he has as little money tied up in material as possible; if he can, he will have the material installed and collected for before he has to pay for it.

3.13 SUBCONTRACTOR'S FAILURE TO PAY FOR MATERIALS

Subcontractors often borrow money on the basis of the money due from the general contractor, and in such case the lender requires checks to the subcontractor to be made out jointly to the subcontractor and to the lending agency. This does not indicate that the subcontractor is in financial trouble, but merely that he has insufficient capital to finance the job himself. In most cases he plans this borrowing when he bids the job. He usually is not anxious to advertise that he is borrowing money, so you should not talk about this fact to others.

The general contractor has a general obligation to pay for all materials and labor on his job, regardless of whether they were ordered by him or by a subcontractor. If the subcontractor goes broke with outstanding bad payroll checks, as often happens, the general contractor must pay these checks. This rule varies with contract provisions and state laws.

Often the subcontractor's reputation is such that suppliers will refuse to furnish him materials on credit. In such cases, the general contractor is asked by the subcontractor to guarantee the payment of materials directly. The contractor is merely agreeing to what he would have to do anyway, but he learns in advance that the subcontractor has insufficient funds to guarantee payment of bills. The contractor will guarantee these bills if he feels that the subcontractor is worth having on the job, as in many cases the only alternative is to get a new subcontractor to finish the work. In any case, this guarantee is the same as if the contractor furnished materials to the job, and the contractor must make payment only to the supplier, not to the subcontractor, to be sure not to pay twice for the material.

Although state laws vary, a contractor generally has the obligation to see that persons who work on his job or who deliver materials to it get paid. If the general contractor pays the subcontractor, but the subcontractor does something else with the money (such as using it to pay older bills from another job), the general contractor will have to pay the bill a second time.

3.14 FAILING SUBCONTRACTORS

A contractor is *insolvent* when he owes more money than he could pay by selling all he owns. This is not an uncommon condition, and firms sometimes operate for years without this condition being known. In construction it is comparatively simple to prepare financial statements that obscure the true condition of the firm, usually by underestimating the cost of current jobs. A firm is *bankrupt* when it requests a federal court to release it from the obligation to pay its debts. A court will do this, providing the manner in which the firm's assets are to be divided up among the creditors. The creditors may request that a firm be declared bankrupt to recover something from what they know to be a bad situation. A corporation is entirely separate from its owners; the corporation may be bankrupt due to the money having been milked off to pay the stockholders or officers, while the owners remain solvent. Consequently, it is preferable to obtain the signatures of owners of a small corporation on a contract rather than the signature for the corporation alone.

A construction company collects and pays out a large amount of money each month in comparison with its net earnings or profits. For the most part, it collects bills for subcontractors and suppliers which it immediately pays. If a company has no money of its own, it may still operate. If it has lost money and is actually insolvent, it is unable to keep up with payments; money collected one month is paid to creditors of the previous month. If it continues to lose money, it will either get further behind or it must increase the amount of incoming money; that is, it can remain behind the same number of days, but the total amount is greater. Thus, a firm may conceal a losing business by continued expansion.

If a firm falls further behind on its debts, creditors immediately take notice and refuse to wait; this will cause the firm to fail. But expansion is not so evident; a firm may continue to operate for some time by continuing to expand, and remain nearly current with its bills; that is, each new job must be larger than the previous one, in order to provide money by overdrawing on the job or by paying late to suppliers so that the firm may remain in business. To do this, the firm must bid low, cutting its profits or increasing its losses even further.

The firm thus falls into a cycle of more jobs, lower bids, greater losses, and more jobs to cover the greater losses. The manager hopes that the next job will be profitable, and sometimes he works out of this situation without anyone knowing what has happened. On the other hand, the contractor who makes a payment to a subcontractor who is in such a situation may find the payment diverted to another job, whereupon the contractor must pay the supplier.

When approving bills of subcontractors, therefore, you should insist on being given the release of liens set up in the contract and be very cautious regarding overpayment to a firm that is expanding rapidly. It is no accident that the failing firm is often the one which puts up the best front—the Cadillac for the manager, the expensive office—as it must impress others with its apparent solvency in order to remain in business.

CONSTRUCTION COSTS

4

Foreman: "Why can you carry only one board when others can carry two?"

Workman: "I reckon they're just too lazy to make two trips!"

4.1 ACCOUNTING METHODS

Accounting in general is a method of finding out if a company is making money. There are many rules for doing this, and an accountant must know them all; to each firm he adapts a set of rules. Each contractor chooses a set he wants to use—sometimes two sets: one for himself and one for the Internal Revenue Service. If a salesperson buys an item wholesale and sells it at a higher price, he can say the difference is his profit. A merchant is not in such a simple position; he knows how much his profit from the sale was, but he must consider his selling expenses and the expenses of his store. He must account for the fixtures which he would not be able to sell at the price he paid for them. A contractor is in a similar position; if he buys a bulldozer for a job, for example, he must know what it is worth at the end of the job or he can't tell what his profit was. He can find out what it's worth by selling it (and on large joint ventures this method is used), but he may need the equipment. He must then have some way to value it.

A contractor has other intangibles, usually lumped as "overhead." He has equipment and salaried people who work on many jobs. He may have worked some inefficient people out of his organization during the job, and after its completion he is in a much better position—he may have cut his costs during the job. This improvement is of value to him only if he continues in business. So at any time, the value of his business depends on future prospects. Rather than try to really figure what his business is worth, he follows the standard procedures of accounting—a set of rules. These rules simplify the work to be done but may not really show the profit-and-loss situation until some time after a job is completed.

4.2 PROFIT OR COST ACCOUNTING

Profit accounting is the determination of overall annual cost. It is done by accountants trained in accounting, but not necessarily in construction. From the results, one pays income taxes and decides to remain in business or to terminate the business and invest the assets elsewhere.

Cost accounting is the determination of costs in detail and in comparison with estimates. The cost reports a superintendent may receive from the office are reports of cost accounting. It is performed by people with detailed knowledge of construction, often termed *cost engineers.* From the results, superintendents and foremen are promoted or discharged, future estimates are changed, and equipment is bought and sold. *Profit accounting is by and for accountants; cost accounting is by and for engineers.*

4.3 TAX ACCOUNTING

Despite the needs of businesses, in the collection of income taxes, the federal government has rules of its own as to how books may be kept. The method of charging overhead costs and of charging use of equipment, for example, may be varied only within certain limits. It is beneficial to the contractor to take maximum advantage of these rules to reduce or to delay his income tax payments. After all, if he himself doesn't know if he made a profit, isn't it expecting too much of him to pay an income tax on these "maybe" profits? Therefore, for certain purposes, he may keep one set of books for tax purposes and another for his own use. This, in itself, is not dishonest; in fact, the Revenue rules specifically allow it. The contractor may want to show a profit to secure performance bonds on his jobs, to keep stockholders happy, or to sell more stock. He also may want to know, for his own purposes, how much the company would sell for if he should die or retire. In this way, he may have three methods of figuring the value of the company:

1. For income tax payments. For the Internal Revenue Service, income and payments are so arranged that taxes are paid as late as possible. In some cases, the tax rate may be reduced by proper accounting management.
2. For surety bond and stockholder reports. A contractor wants his business to appear as profitable as possible. Overhead is charged against later jobs, for example, and equipment is shown at a higher value than for tax accounting.
3. For management use. Accurate up-to-date costs are needed so that the value of foremen, superintendents, and project managers can be found while there is still time to promote or fire them.

No matter how books are kept, the amounts paid out and received are the same; consequently, *eventually* all methods will show the same profit or loss, but it may be necessary to sell out the business to bring them together.

4.4 CASH ACCOUNTING

The most important question of accounting to a contractor may be whether or not he has the cash to pay current bills to maintain his credit standing. There is considerable truth in the old saying that banks will lend only to people who can prove they don't need the money; and other firms will provide credit only to firms who can prove they don't need it.

When bidding a job, a contractor plans his cash requirements. His sources of money are his own cash and anticipated receipts, the amount he may borrow from banks or others, and the demand for money as governed by the credit he may obtain from subcontractors and suppliers. He may make errors, particularly in estimating receipts, which may put him in a situation where he has no cash but is operating at a profit and is basically solvent. Sometimes his bonding company may lend him money under such conditions, as it is the cheapest way for it to complete bonded jobs.

You will soon find out which are weak and which are strong firms, considered on the basis of cash. This strength or weakness may have no relationship to ability or even the value of their business. You will get the best bids from weaker subcontractors by keeping them paid up to date; stronger firms may not particularly care when they are paid. Approval of amounts for subcontractors should be made with this in mind, and the contractor may change the amounts you have approved for payment for this reason.

4.5 LABOR-COST ITEMS

An estimator, project manager, or office engineer designates the cost account system and the code numbers to which either he or the bookkeeper charges the material and subcontractors' bills. Similarly, an estimator prepares a list of labor cost items from the estimate; all labor is charged to one of these items.

The superintendent is furnished a list of such items, which may be as few as half a dozen or as many as several hundred. The list may consist of names of items or it may have a code letter or number, or a combination of numbers and letters, which the foreman or timekeeper puts on the daily or weekly payroll. The amount of detail required depends on the type of estimating used and on the size of the job. For example, you are building a warehouse with concrete exterior foundation walls and concrete continuous footing, and concrete interior footings with piers. You must lay out the footings, dig the holes, set steel, pour concrete, form the foundation walls, and strip the forms. You may be required to rub the foundation walls. The simplest breakdown would be merely a statement:

Concrete foundations 150 cu. yds. $932.00

Usually, but not always, the quantities of work are shown in the breakdown, and some firms require daily reports of work done in terms of quantities of work in place.

The items above may be elaborately coded, as illustrated in Fig. 2. The numbers shown in Fig. 2 illustrate a system frequently used when accountants prepare the numbering system. Totals are at the top for each classification, and are underlined. Sometimes several subtotals are used, a practice that can be very confusing. The same numbering system may be used for materials and labor, in which case 1.51 might be formwork and 1.52 the materials for formwork; in this case, the superintendent does not use many of the numbers at all, since they are for material only.

Code	Item	Quantity	Amount
1.	Concrete Foundations	150 c.y.	$1884.00
1.1	Building layout	L.S.	100.00
1.2	Excavation		
1.21	Wall footing excavation	50 c.y.	100.00
1.22	Interior footing excavation	50 c.y.	100.00
1.3	Reinforcing steel		
1.31	Wall footing bars	300 lbs	20.00
1.32	Int. footing bars	300 lbs	24.00
1.4	Concrete placement	150 c.y.	600.00
1.5	Foundation forms		
1.51	Wall foundation forms	1,300 sf	600.00
1.52	Interior piers	50 ea	200.00
1.6	Finish exposed concrete	650 sf	140.00

Figure 2

Some of the numbering systems in common use are:

1. Letters and numbers, such as *C-1, C-2, C-3,* etc. The *C* usually is an abbreviation for some category of work, such as *concrete.*
2. Abbreviations, with no further breakdown, such as *Conc, Br, Lumber, Millwk,* etc.
3. Consecutive or semiconsecutive numbering systems. With this system, numbers, reported on payroll slips are merely *1, 2, 3, 4,* etc.
4. Numbers without decimals, such as *201* and *212,* the first digit representing a classification of work or trade.
5. Construction Specifications Institute numbering used in most specifications.

You will probably have no choice as to the numbering system to use; if you do, there is no reason to use one any more complicated than simple numbers, such as *1, 2, 3, 4.* This may make a little more work for the bookkeepers, who will have to convert some of these results into other numbers in the office, but this work is small, and you are much less likely to make errors in the field than if a more complicated system is used.

4.6 WORK-ITEM DESCRIPTIONS

Since the work items must be selected from the estimate, the superintendent has little opportunity to choose what they are to be. But some contractors do not keep cost accounts by a labor-cost breakdown at all, or do it only by very large items, such as all framing lumber or all concrete work in one item. Others attempt to break down labor costs to such small items that it becomes quite laborious. On a large job with considerable repetition of formwork, for example, it is helpful to know how much money is spent on cutting lumber for forms, making forms, setting them, stripping, and resetting. On a small job, this detailed breakdown is impractcal.

Even if the contractor does not require that a labor-cost breakdown be made, the superintendent needs this information for his own use. The superintendent can do this in as fine detail as his available help can handle; if he has a payroll clerk with little to do, the clerk can take care of an elaborate system of cost controls. If the clerk is overloaded, or if the superintendent is alone, he can do very little. For example, the superintendent may use large cost items such as complete floors, which take half his labor force as much as a week to complete, if he must do his figuring on the back of a note page. If a clerk can handle it, the cost items noted may be so small that one worker-day is sufficient to complete such a small work item. The superintendent may not even keep any written records, but simply remember how many worker-days certain work takes; later he can use this information when he makes an estimate for his own purposes. Rather than resisting clerical work of keeping labor-cost accounts, the superintendent is usually anxious to learn what different kinds of work cost—not only that he may work as an estimator, but also in case he wants to start his own business.

The contractor may conceal all cost information from the superintendent; the superintendent may not even be allowed to see the labor costs after they have been reported as worker-hours. The superintendent can always figure his costs, at least roughly; whether this is more or less than the estimate is then a matter for the contractor to worry about.

Unfortunately, work items are not always clearly described in the list furnished to the superintendent. You must then look through the plans and classify all the work you know has to be done in accordance with your list. If you have curb angles to set in concrete forms, for example, and there is no work item for it, you can expect this item to be included in the formwork. If you have anchor slots to set in columns, this will be included in column formwork if there is no special item provided.

There are a number of items that are always vague. If an item for *Material handling* or *Hauling* appears, does it refer to *all* hauling? Is all formwork hauling included? What does *Cleaning* mean—general cleanup, cleaning forms, cleaning floors, sweeping, cleanup after subcontractors, moving scrap to the dump? If possible, try to clarify all such doubtful items. Unfortunately, this is not always possible. The estimator may have been reporting such items for years, and he doesn't really know how things have been charged. The same superintendent charges things the same way, and the estimator uses these reported costs for the next such item in his estimate. In this case, the best source is the more experienced superintendent. If an amount of money is budgeted to the superintendent for such items, the superintendent will make the reported work match the money and then charge the rest to something else. There is a strong tendency for field people to attempt to make the estimate come out right, as they know that if it doesn't, the office people will be unhappy. In fact, if an item appears on the work item list which is entirely ridiculous or isn't even on the job, it is quite possible that a foreman or time clerk will utilize the item to charge time to; he sees no point in being high on one item when there is an unused item that *must* have been done.

4.7 ACCURACY OF LABOR COSTS

Labor costs are notoriously inaccurate. Some large firms have given up entirely any attempt to determine what their detailed costs are. Others have put special personnel in the field to see that such items are reported properly. Another approach is to use written work orders, describing the work for each such cost item, and to prevent the foreman from beginning the work until he has a detailed description of the work in his hand.

Unless you the superintendent give your personal attention to the reporting of costs, they are almost sure to be improperly reported. Some of the reasons for this are:

1. Descriptions are vague, as mentioned above. The superintendent himself may not know what is supposed to be included in a cost item; if he writes a description in more detail, he may not agree with someone else. So he takes no chances, and hands out lists just as he gets them.
2. Responsibilities are separated, so that the superintendent or foreman is responsible for getting the work done, and another employee, often responsible to the bookkeeping section, is responsible for reporting cost breakdowns and sometimes for reporting time for payrolls as well. This procedure is usually an attempt to correct previously poor cost reports made by foremen.
3. Construction supervisors have had little experience with paperwork of any kind, and feel their job is to put work in place, not write about it; they believe that the numbers they write on a piece of paper do not change the cost or increase the profit on their job.
4. The foreman has little confidence that the costs reported will be to his advantage and fears that cost reports will be used to discredit his work; he feels especially vulnerable to criticism because he doesn't know what is expected or what his own performance actually is. He feels that if he is doing poorly, he will be laid off, but if he is doing well he will not be informed of that fact.

Preventing all these situations rests on your shoulders. If descriptions are inadequate, ask the contractor about them. If you must decide yourself what is meant, do so. If you state clearly what you are including in a particular work item, the estimator can usually adjust the estimate to compare with your item. A great deal depends on the personal relationships involved—between you, the estimator, and the contractor. If the estimator feels his position is endangered by questions, and the contractor is not particularly interested in the matter, there is very little you can do. The reported costs are only as good as the estimator who uses them; if he already feels he knows what the costs are, he is not likely to change his mind anyway. On the other hand, if the estimator is anxious to cooperate, your questions may cause an improvement in everyone's cost reporting.

Once the responsibility for cost reporting has been taken out of the superintendent's hands entirely, it can be changed only if you are careful to make sure that the time clerk on the job is properly reporting time. First, be sure you know whether the reporting is being done consistently, and then learn enough about it so that you can supervise it if necessary. Since the time clerk may be a person not subject to your supervision, there is no need to interfere, whether or not the reports that are going in are correct. If it is being done well, there is no reason to question it; if it is not being done well, it will land in your lap soon enough without your pointing out to anyone where the trouble is. You will find that you will get into enough trouble (and extra work) merely by answering truthfully to questions asked you; there is no necessity to look for trouble.

The third difficulty—lack of interest on the part of foremen—is one that you must overcome by explanation. It *does* make a difference if the costs are currently and properly known. Foremen who have been in business themselves realize this; others need to have the matter explained. By reporting their own work accurately, they not only help the company get more—and profitable—work, but if their costs on a particular type of work are low, they assure themselves of a continuous job and perhaps bonuses. Bonuses can best be paid not for observed performance, but on the basis of proven performance—and the cost records are the way to prove it. Thus, these costs are your weapon for extracting higher earnings from your supervisor, both for yourself and for your foremen. You will find that most contractors are generous with savings, especially if they haven't made them yet! In spite of his apparent concern with wage rates, a contractor's real concern is with overall costs, and if you can show him this is low, higher earnings are easier to obtain. If you get bonuses for your foremen based on costs, you increase your influence on the foremen; the contractor is obligated to raise your earnings, both because of your efficiency and because of your influence on personnel he knows to be efficient.

The fourth objection, that costs reported by foremen will be used against them, is more difficult to combat. If all cost records are kept on the job, the foreman is available to discuss the reports and he feels they are his own. He takes pride in low costs not just because he knows they are low, but because the employer appreciates them. This is a powerful argument for stating estimated amounts on copies of the cost items available to the foremen, and for keeping him informed of current standing. On the other hand, errors on the part of the estimator may unduly disturb the foreman; that is, if the estimated amount on an item or on several items is too low; the foreman will get discouraged and may quit; for he won't know if the estimate is low or not, but he will know that his work is considered unsatisfactory.

The fact is that these cost returns *will* be used to determine the efficiency of the foreman, but they should be used only by you, his immediate supervisor. You know the circumstances under which he is working and can decide if the estimate is reasonable. You can discuss the item with him and can transfer him to other work where he may be more efficient. If production is lower than the estimate on some items, this is reason enough to change foremen—not to get rid of the foreman, but to find out if the estimate is right. If *no one* can do the work for the estimate, it would appear that the estimate is too low. These costs are also the check the contractor is using against you—you should have the responsibility to decide if the work is being done the best way.

Keeping cost records on the job keeps them in the hands of the superintendent, and more—the foremen know that the home office can't check on individuals. The traditional authority of the superintendent to judge his own personnel is rarely challenged in other matters, and it should not be modified because of cost accounting methods. As one superintendent said when asked by the contractor, "How is Smith doing?", . . . *"He's still working for me."* That is, the superintendent expressed his opinion of his subordinates by keeping them or firing them; if he still had them, they were the best available, and a detailed discussion of their virtues or faults was useless.

4.8 SPECIAL COST REPORTS

Some reports are of special importance, well beyond their apparent reported amounts. If the firm lacks cost records of a certain kind, the unit cost of a very small item may be blown out of all proportion on another estimate. For example, a house contractor has never built a concrete slab structure. He contracts a house with a concrete slab roof on the front porch—a small quantity, but enough to indicate the cost involved per square foot. He uses this cost to bid a building with a concrete structural floor and roof, 20 times as large as the porch. Accurate cost reporting is essential on the porch roof since the amounts obtained are to be magnified 20 times. Of course, the larger area will have a cheaper unit price (probably), but a base figure is helpful.

In other cases, a subcontractor may be available on, say, all concrete slab work on grade. A job is selected to make a cost determination of the work as being done by the general contractor, but the time spent by carpenters making up partitions ahead of the concrete floor is reported in the concrete slab work. This makes the concrete slab work appear higher than it really is, and the contractor concludes that the work should be sublet. The framing work continues to be reported on each job, but there is no way now to determine that the cost is greater—costs are no longer being reported for the slab work, and as long as this situation continues, the contractor does not discover that he is doing the work the expensive way.

4.9 COST MISCHARGES

If charges are made to an item not listed exactly the same way in the labor cost breakdown as in the time report, the person working up the totals can't tell what to do with it. For example, if the list says *Footings* and you report a cost on *Interior footing,* a clerk who knows nothing of the job is confused. He doesn't know that this item is a part of the overall footings item. Even a small discrepancy of this sort makes the clerk—and the estimator—doubtful about the accuracy of the whole report, and may cause them to discard a report which is substantially correct. If your office people are slow about their work, you may get calls about items of this sort a week or two in the past—usually too far back to remember.

If your own clerk makes up cost reports, be sure they are done promptly to avoid reliance on memory of weeks ago. One way is to require Saturday work if the records of the week—usually through Wednesday or Thursday—are not completed. Overtime can sometimes be avoided by authorizing all necessary overtime to complete the work—and pointing out that if the clerk can't do the work in 40 hours, you will find one who can.

When numbers are used to designate labor-cost items, there is little chance of confusing items unless the foreman's printing is illegible. If he takes the same care with these numbers as he knows he must with time reported, they should be clear. If the decimal system of designating items is used, the overall heading number should not be given to the foreman, or he will use the heading number rather than the proper detail number. For example, if *135.00* is the total for all footing work and *135.10* is for excavation included in footings, time can be charged only to the number for excavation (*135.10*). There must be other numbers for other items. If time is charged to the total (*135.00*), there is no way to break down *any* of the detailed items, since you can't find out what part of the work was erroneously charged. Sometimes labor-cost items are too small; the amount of these small items is in error, but the larger cost items are still correct. In the example above, if some time from *135.20* were charged to *135.10,* neither would be right, but the total of the two (*135.00*) would be still correct.

4.10 CHANGES AND EXTRA COSTS

When a change in the work is authorized by the architect, it is usually necessary to keep an accurate account of the time to complete the added work. Some contractors keep no breakdown of labor on the main contract at all, but they must keep an account for either extra work or change orders. A *change order* is a direction from the architect to do work not in the contract; *extra work* refers to the actual work done in addition to the contract. Extra work is often work which the contractor believes to be additional to the contract but which the architect may not allow. There is no doubt about a change order; both sides agree that it is extra work. But there may be extra work done before a change order is given, or extra work may be claimed by the contractor but not paid by the owner. In these cases, costs must be kept as extras, although they may be eventually absorbed by the contractor as *estimate* errors and omissions.

In each case, it is essential that the cost of a change be kept separate, as this cost may be used to determine the contractor's payment for the work. In private work, extras are often authorized on a cost-plus-percentage basis. The superintendent may use a different form of payroll report for such extras, largely because the architect must be assured that the records are accurate. Consequently, the separate form bears the signature of the foreman or superintendent for each charge, and of the owner's inspector or engineer also, if there is one on the job.

The U.S. government is forbidden by law to pay for extra work on a cost-plus-percentage basis. Payment is for the value of the work, but this may be determined after the work is completed; one way of determining the value is to use the actual cost of the work as a base. In effect, therefore, the item is settled on a cost-plus basis due to the failure of the parties to agree on the price before the work is completed. Sometimes the contract makes the government responsible for determining the value of extra work, but often there are no persons in the government organization capable of doing so, and in any case, using the contractor's estimate is customary.

4.11 WORK-ORDER SYSTEM

Work order, as used in construction, usually refers only to extra work authorizations. Some contractors call their labor-cost items work orders and write a work order for each item for which separate costs are desired. The term has been borrowed from shop work; nearly all shops, including naval shipyards, use such a system. Shops which work on items of work that move from one area to another, such as repair shops and stone milling shops, may call the work order a *job ticket.*

The work order (Fig. 3) is a written order given to the foreman by the superintendent. It may be written by the project manager or by a scheduling engineer. It substitutes written for oral instructions and information. It is a superintendent's tool, as well as a method to assure that the labor costs are properly charged. The simplest case is merely a

Work Order No. 12

Date 2-6 Job Hospital

Foreman Yohannon Trade Carp. Lab.

Formwork for second floor and beams not including columns below.

Strip first floor formwork, clean, repair, make new formwork as necessary, install in place for second floor pour. Includes pans, slabs, beams, stair openings (but not stairs), curbs, makeup necessary for this floor. Includes setting inserts of all kinds, hatch frames, and anchor bolts except those for equipment in machinery room which will be set by mechanical contractor. Includes watcher during pouring, but no stripping of second floor.

Begin 3-5 End 3-13

Est MD 30

Material--Plywood, PO 43, Jones Co.
--Exp. Smith.
Nails--PO 23, Newton
--Exp. Smith.
Scaffolding--PO 24, Patent,
--Exp. Jones.

Plans: Ch. rev. 1-6 to sh. 3

Supts. Notes:
See Newburg (Saw man) for all material, Johnson for crane service.

Figure 3 Work order.

note from the superintendent to the foreman, outlining the work included in a labor-cost item. The superintendent may want to include other instructions, and as he does so, the work order becomes more elaborate.

A work order intended to give the foreman full information is illustrated. Most of this information is given to the foreman orally when the job order is not used. Some of the information will not be needed by the foreman unless material is misplaced or fails to arrive. The job order is made from planning, drawing, and material delivery information, and consequently should be made after all data to complete the work item are available. If these orders are made up by the project manager or by other persons in the office, they are given to the superintendent for distribution.

The superintendent may hold up the work by keeping the orders, may assign them to available foremen, or may change the instructions given on the sheet.

Preferably, a work order is given to one foreman; if the work requires two crews, it should be separated into two items. The two foremen concerned are then responsible for the cost of their items and can't complain about other trades' charges. In the example, the definite operation of moving and resetting forms is described; no attempt is made to separate repair from resetting forms. If these were separated, the foreman would be encouraged to keep repair costs down by taking too much time in stripping, or to keep stripping costs down by increasing the speed of stripping and increasing repair costs. One person should be responsible for both.

In the example, the foreman is cautioned that some work he normally does is to be done by a subcontractor; if measurements were furnished by a subcontractor, the foreman should know where to get these dimensions. The item is ended at a definite time—here, when the concrete floor is poured. At this time, there are only one or two people working on this item, so there is less likelihood that a worker will be carried a day too long on the item.

Items in the material section of the work order tell the foreman where the material should be. If it is missing, he can check the shipper, whose name is given; he can find the purchase order copy in the superintendent's office, or he may call the expediter (here abbreviated "exp.") to see where the order is. In this instance, the saw man has been designated to be responsible for all form material by a superintendent's note to that effect.

If the form foreman is responsible for cost of equipment he uses (that is, if the equipment cost is charged to the foreman's cost item), the foreman will be unlikely to tie up equipment, and must choose whether to use a crane or to use labor for some items. The note "Johnson for crane service" indicates to the foreman that he is to request Johnson (the operating engineer or master mechanic) to schedule the crane. If the form foreman is responsible for cost of equipment, the equipment operator uses the number given him by the foreman for his own cost report, just as though the operator were another contractor. The work-order form may be for notations regarding plan changes, and when changes are coming through rapidly, the work can be held up by keeping the orders in the superintendent's hands until revisions are completed. When revised plans are issued, an order accompanies the new plans. This may be a revision of the old order, and it may be required that the foreman turn in the old order, as he is required to turn in old plans by many contractors. This is to ensure that obsolete plans are not used for construction.

The scheduled beginning and ending dates are set as described in Chapter 5, and may be shown on the work order.

4.12 WORK ORDERS FOR EACH JOB

Few contractors use the job order with all the details previously noted, on parts of a job. More often, small contractors use a similar order for small jobs, calling it a *job order,* as shown on Fig. 4. Small jobs may have an order that replaces plans and purchase orders; it may also become the invoice and time report. Work or job orders may be used for any of the purposes mentioned; one large contractor uses job orders for planning operations, such as pouring the concrete floor of a building, but does not use this item for cost accounting. A separate breakdown is used for labor-cost reporting.

The basic unit of job planning may be a job order, work activity (used when planning by the critical-path method; Section 5.27), or cost account number (for cost accounting). Both of the forms illustrated are the size

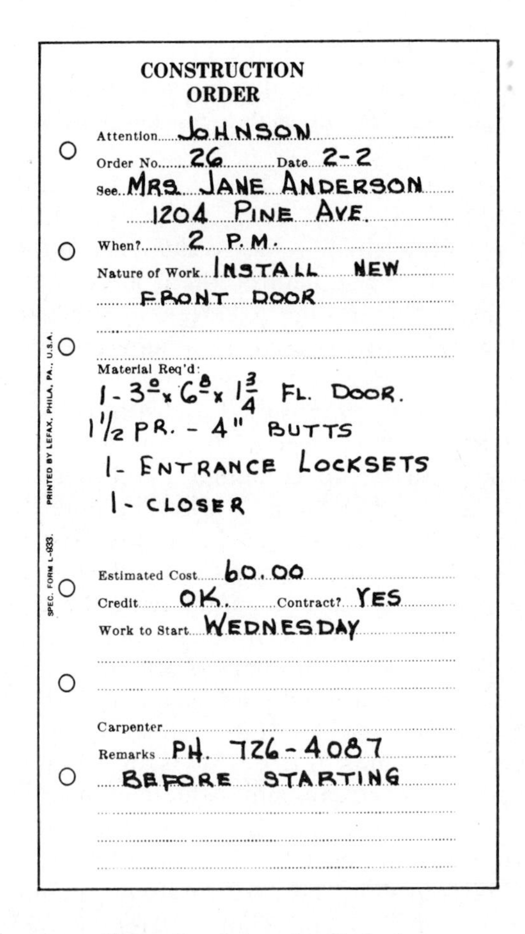

CONSTRUCTION ORDER

Attention JOHNSON

Order No. 26 Date 2-2

See MRS. JANE ANDERSON

1204 PINE AVE.

When? 2 P.M.

Nature of Work INSTALL NEW FRONT DOOR

Material Req'd:

$1 - 3^0 \times 6^8 \times 1\frac{3}{4}$ FL. DOOR.

$1\frac{1}{2}$ PR. - 4" BUTTS

1- ENTRANCE LOCKSETS

1- CLOSER

Estimated Cost 60.00

Credit OK. Contract? YES

Work to Start WEDNESDAY

Carpenter

Remarks PH. 726-4087 BEFORE STARTING

SPEC. FORM L-833. PRINTED BY LEFAX, PHILA., PA., U.S.A.

Figure 4 Job order (Lefax).

of a pocket looseleaf notebook; more often a full-size sheet is used on a clipboard.

The *daily time report* combines all the work orders for one day on one sheet. A daily report enables the payroll clerk to enter the hours each day, somewhat reducing the work to be done at the end of the week. Also, if an independent check of the workers' attendance is being made, this may be checked daily. The weekly report is more often used by small contractors, for whom preparation of the weekly payroll requires only a short time.

Current—usually weekly—reports are also needed to check the progress of the work and to see if changes in supervision—either by changing operations or people—have to be made. The contractor watches these reports very closely.

Regular cost reports will make foremen more cost-conscious. It is a principle of management that any effort, even poorly made, to change conditions will improve efficiency. In a series of experiments, lighting was increased in a factory, and the manager was happy to see production go up; then when lighting was reduced back to the old level, production rose again! Men produce more if they *care*—and they care only if it appears that the foreman and the superintendent care.

A foreman who reports on his production or helps the superintendent or cost engineer make out the report will take more interest in production than if no report is made or if the foreman doesn't see it. Also, the report must be personal—it must be of one foreman's output. A worker is not at all interested in the total output of his crew and another's, but is concerned about how his crew compares with another.

The cost report, therefore, does more than show where production is low—it prevents it from becoming low in the first place. Needless to say, you can't fool the foreman about what you are doing; the trades have been "worked" this way for a century and you can attain production only with the foreman's full cooperation.

4.13 COST REPORTING FORMS

Payroll reports from the job *must* provide information for paying the workers, and *usually* provide information on how much time was used on each part of the work. These two kinds of information may be provided on one report or on two separate ones. In any case, the labor-hours should be checked from each form against the other; if labor cost is calculated on the job, the forms should agree on the total cost. These reports may be made daily or weekly.

Two methods of reporting both pay and cost breakdown information on a single report are shown on Figs. 5 and 6. The Fig. 5 report requires a

SPEC. FORM L-817 PRINTED BY LEFAX PHILA., PA., U.S.A

Order No. 12 Time, Week Ending MARCH 10 19

NAME		Sun.	Mon	Tue.	Wed.	Thur.	Fri.	Sat.	Total Paid Hours	Rate	Total Amount
JONES, H. B.	Straight Time, Hrs.		8	8	8	8	8		40	3^{80}	
	Overtime, Hours										
WYMAN, PAUL.	Straight Time, Hrs.		8	–	8	8	8		32	3^{50}	
	Overtime, Hours										
JOHNSON, P. P.	Straight Time, Hrs.		8	8	8	8	6		38	3^{50}	
	Overtime, Hours		2						2-OT		
NEWARK, M. N.	Straight Time, Hrs.		8	–	–	–	–		8	3^{50}	
	Overtime, Hours										
JAMISON, O. A.	Straight Time, Hrs.		–	8	8	8	8		32	2^{60}	
	Overtime, Hours										
PROVINSKY, A	Straight Time, Hrs.		8	8	8	8	8		40	1^{50}	
	Overtime, Hours		2						2-OT		

Figure 5 Daily time report—This type of report is made out daily for each work item, and is used when work items each include several days' work for a crew (Lefax).

separate sheet for each work order, cost classification, or part of the job; these three terms mean the same for cost-report purposes.

4.14 EQUIPMENT REPORTS

Operating time and operating costs for construction equipment are reported in a wide variety of ways to different contractors; some do not ask for job reports at all.

If equipment time is to be charged to one cost account, a single weekly equipment report such as that shown in Fig. 7 may show several pieces of equipment. Figure 8 shows a report for operating time and payroll time for personnel operating or working with the machine. It does not allow for fuel and other operating costs. The latter costs may often be obtained from invoices, if the contractor wants them. A similar form is often used by firms who rent equipment by the hour or day. It is submitted to the customer-contractor to back up invoices, or may be used as an invoice.

4.15 COST-STATUS REPORTS

Regular cost reports are of two types—*status reports,* which show how your cost stands to date, and *period reports,* which show the cost of work

DAILY TIME REPORT

DATE *FEB 2, 19—*
FOREMAN *Perry*
TRADE *Carp.*

NAME	WORK ORDER NO.			
	1	3	6	21
Johnson, P.	8			
Smith, D.M.		8		
Fortuna, I			4	4
Halperin, D.	2			6
Royer, P.	1	1	6	
Reeves, W.L.		2	2	

SPEC. FORM L-1640-A PRINTED BY LEFAX, PHILA. 7, PA. U.S.A.

Figure 6 Daily time report—This report is made out daily by each foreman. It has the advantage of fitting in a pocket-size notebook (Lefax).

done in a particular day or week. Either or both may be used on any one job.

Large construction firms usually have these reports made by computer, and the superintendent gets a printout report. Small firms often make no reports at all.

A status report is a list of cost items and amounts, with the work done to date compared with the cost to date, as shown in Fig. 9. This illustration is much simpler than reports in common use. The second floor, for example, may be broken down into different types and operations of formwork, placing of steel, and pouring and curing of concrete.

The estimated percentages completed, or quantities completed, may be made by the superintendent, foremen, or on larger jobs, by a cost engineer. They may be made from looking at the work or from measurement of quantities of work in place. The "cost to date" is obtained from

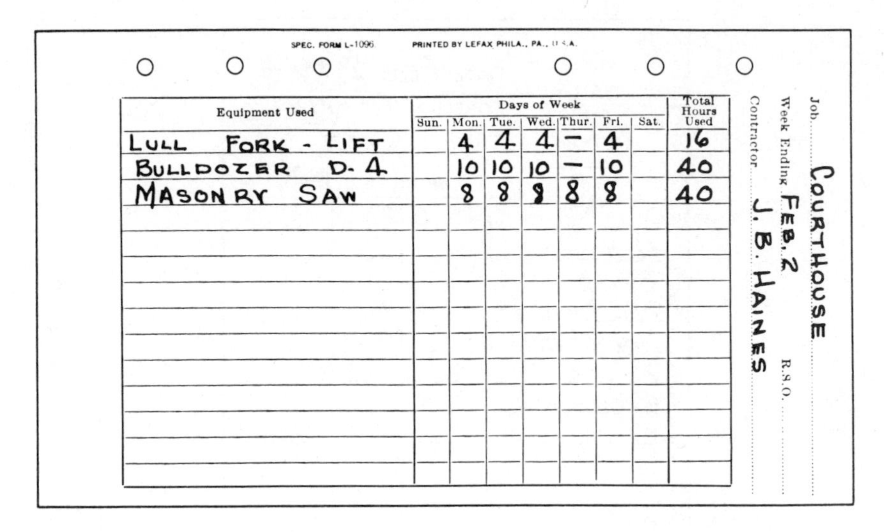

SPEC. FORM L-1096 PRINTED BY LEFAX PHILA., PA., U.S.A.

Equipment Used	Sun.	Mon.	Tue.	Wed.	Thur.	Fri.	Sat.	Total Hours Used
LULL FORK - LIFT		4	4	4	–	4		16
BULLDOZER D-4		10	10	10	–	10		40
MASONRY SAW		8	8	8	8	8		40

Job COURTHOUSE

Week Ending FEB. 2 R.S.O.

Contractor J. B. HAINES

Figure 7 Equipment report—This report serves as a cost report for several items of equipment (Lefax).

the records of reported labor costs, and these may be kept in the home office. Consequently, you will usually report only the percentage of work completed.

If the contractor uses these reports to judge your production, as he will if they are any good at all, you should review the completed reports; there are usually good reasons for poor production on certain items, and sometimes there may be outright mistakes. Since many contractors do not like to return any cost reports to the field office, it is to your advantage to compute costs on the job if possible.

When quantities are used as a basis for estimating percentage completion (or unit prices are used for estimating value of work in place, which amounts to the same thing), serious errors can occur due to difference in actual labor costs. Formwork, for example, is costly at first per square foot because of the makeup cost required. Excavation cost varies considerably with hole size. Sometimes a unit-cost method makes you look good, sometimes bad. If your costs look good, don't let the contractor rely on them too much; the more severe will be his reaction when he finds out you—or your costs—are not so good as he expected.

Status reports a week apart can be compared to see how well you have done during the week. Such a comparison is usually not reliable, however, since one report may be estimated too high and the next too low. The weeks' difference will then be affected by both errors. For longer periods, such errors are not significant. It is traditional that early rough work usually shows a gain and later finish work a loss—there is a natural tendency to overlook the detailed labor necessary in finishing operations.

DAILY EQUIPMENT REPORT

DATE FEB. 2, 19—

OPERATOR HANSEN, O. B.

EQUIPMENT D-4 Bulldozer

	WORK ORDER NO.						
	1	3	8	10			
OPERATOR:	2	4	6	1			
LABORER:							
LINCOLN, G.	2	4	6				

MACHINE OPERATING TIME

12 HOURS

Remarks: (Repairs, Adjustments, Lubrication, Etc.)

TRACK ADJUSTMENT
NOW CLEAR OUT

SEEC. FORM L-1650. PRINTED BY LEFAX, PHILA 7, PA., U.S.A.

Figure 8 Daily equipment report—This report serves both for equipment accounting and as a payroll form for the operator and his assistants (Lefax).

Cost Status Report

Item	Budget Amount	% Complete	Value in Place	Cost to Date	+ or -
Concrete Work					
1. Footings	$ 250	100	250	275	-25
2. Slabs	500	75	375	500	-125
3. Columns	200	25	50	40	+10
4. Second floor	1500	–	–	–	–
Totals	2450		675	815	-140

Figure 9 Cost status report—A report like this shows if labor cost is within the estimate.

4.16 PERIOD LABOR REPORTS

Many reports are made daily. Masonry and concrete placing can be readily reported by quantities completed per day, as there are few operations completed in a day or two. Some firms require daily cost reports on all work; probably some do so to keep the foreman careful of costs rather than for accurate results.

Daily reports of work completed *can* be easier than weekly reports. At the end of a day, the starting point for the day can still be remembered. In the case of masonry, a count of quantity laid can be made daily from the color of masonry joints.

Figure 10 is a sample labor report of concreting operations. This form was first used in 1910, when all concrete was job-mixed, and it is still in use. The form shows the unit cost, which is not as useful for larger jobs

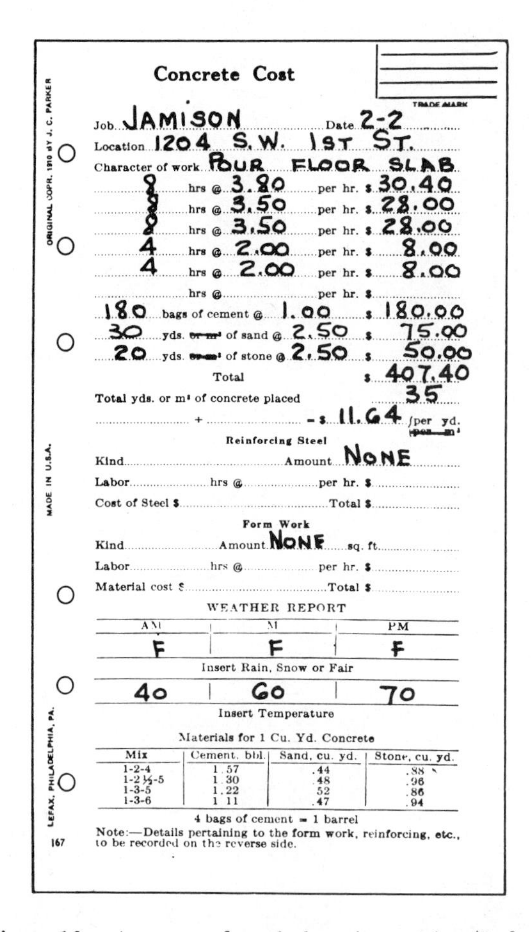

ORIGINAL COPR. 1910 BY J. C. PARKER

Concrete Cost

TRADE MARK

Job JAMISON Date 2-2
Location 1204 S. W. 1ST ST.
Character of work POUR FLOOR SLAB
8 hrs @ 3.80 per hr. $ 30.40
8 hrs @ 3.50 per hr. $ 28.00
8 hrs @ 3.50 per hr. $ 28.00
4 hrs @ 2.00 per hr. $ 8.00
4 hrs @ 2.00 per hr. $ 8.00
hrs @ per hr. $
180 bags of cement @ 1.00 $ 180.00
30 yds. ~~or m³~~ of sand @ 2.50 $ 75.00
20 yds. ~~or m³~~ of stone @ 2.50 $ 50.00
Total $ 407.40
Total yds. or m³ of concrete placed 35
\+ = $ 11.64 /per yd. ~~/per m³~~

Reinforcing Steel

Kind Amount NONE
Labor hrs @ per hr. $
Cost of Steel $ Total $

Form Work

Kind Amount NONE sq. ft.
Labor hrs @ per hr. $
Material cost $ Total $

WEATHER REPORT

AM	M	PM
F	F	F

Insert Rain, Snow or Fair

40	60	70

Insert Temperature

Materials for 1 Cu. Yd. Concrete

Mix	Cement, bbl.	Sand, cu. yd.	Stone, cu. yd.
1-2-4	1.57	.44	.88
1-2½-5	1.30	.48	.96
1-3-5	1.22	.52	.86
1-3-6	1.11	.47	.94

4 bags of cement = 1 barrel

Note:—Details pertaining to the form work, reinforcing, etc., to be recorded on the reverse side.

MADE IN U.S.A.

LEFAX, PHILADELPHIA, PA.

167

Figure 10 A report of work done in one day (Lefax).

as is the actual money cost gained or lost during the day. If unit costs are low for 4 days and high 1 day during a week, the net difference may still be a loss if the quantity was high on the high-cost day.

The masonry-cost report form (Fig. 11) shows the value of work in place, according to estimate unit prices, and compares this with the day's cost of wages. The difference represents the loss or gain for the day, and can be added up day by day to figure an overall gain or loss for the job to date.

Note that in both cases above, the cost report is kept separate from the payroll report. If there is an error, therefore, it will not be found by the payroll clerk. This type of daily report by trades has the disadvantage that some workers may be omitted from the cost report. There is less chance for error if the cost report total corresponds to a cost item on the payroll cost report so that they can be checked against each other.

Daily cost reports are most reliable when the foreman making out the reports receives a bonus based on what he saves under the estimate. There must be an overall check, of course, but a foreman who can barely add two and two will often show an amazing increase in arithmetical ability

Masonry Cost Report

Job SMOTHERS Date 2/4/80

Work put in place - -

500	8" C.B.	@	.30	$150.00
1,000	12" C.B.	@	.25	250.00
3,000	Brick	@	.20	600.00
		Misc.		60.00
		Total		$1,060.00

Cost of work done - -

60	hrs.	@	12.00	$720.00
60	hrs.	@	6.00	360.00
8	hrs.	@	13.00	104.00
		Total Cost		$1,184.00

Ahead

Behind $124.00

H. C. Tate

Figure 11 Masonry cost report—This report shows one trade's gain or loss for one day.

and understanding when his own money is involved! When such an arrangement is made, it will be impossible to cut the foreman's percentage on another job, should the estimate turn out to be too high, unless the foreman realizes this situation beforehand.

In all reports with numbers for dates, be sure you are using modern United States practice of the month first, for example 2/5/80 for February 5, 1980. In foreign countries and occasionally in the United States, 2/5/80 is written to mean May 2, 1980.

4.17 SPOT-COST REPORTS

Occasionally, you will have a type of work which is too small to carry as a separate cost item, but which may be important. For example, in laying small precast lintels or other precast work in brick walls, the time required for one lintel is too small to report separately; nevertheless, you may need to know the unit cost. You can make several checks on single lintels being placed; if the office shows an interest in these pieces of information, send them in as a note of "spot" costs; if not, keep it yourself.

Another example is the cost difference in laying block at platform level, at waist-high elevation, and at shoulder height. It is generally recognized that moving scaffold, which keeps bricklayers at their most comfortable height, is more efficient than that which requires moving upward in stages; but how much more efficient? By checking the time for laying each course, you can estimate the increase in production expected if all courses were waist-high, as with movable scaffolding.

Spot costs furnish labor-hours per unit under observed conditions, and must be adjusted for time lost. They are therefore valuable for estimating.

4.18 PIECE RATES

Although payment for labor on the basis of quantity of work done, *piece rates,* are common in most of the country on housing, it is rare on union and commercial work. However, it is always useful to consider payment in this way. By observing the production of various workers for short periods of time, you can find out how much work they might do under a *piece-work payment* plan. This is usually paid as a guaranteed minimum, with a bonus for work over the set standard, so that the total payment is for a certain amount per piece. In general, you should expect the slower workers to catch up with the faster ones, in order to earn a day's pay; to the extent that they speed up (never use *that* expression, though!— as it has been used in a derogatory sense by labor unions), they should get more pay. As a rule of thumb, personnel working under a piecework plan should earn one-third to one-half more than the going hourly rate. Don't expect the better workers to work any faster than the rate at which you observe them, although they may increase production by wasting less time. You may find that even the better workers will produce at a daily rate of one-third less than the rate at which they work for short periods.

For example, suppose that you observe workers setting windows, all under similar conditions, varying from 30 minutes to 50 minutes for each window per worker. You may find that the 30-minute worker sets 13 windows per day and the 50-minute worker correspondingly fewer. You are paying $100 per person per day, and averaging 8 per day, or $12.50 per window. The lowest cost is $7.70 per window.

Prior to setting a piece rate, you should bring the lower producers up to the proposed rate, by increased supervision, layoff, or transfer of workers. The piece rate will then be an immediate increase for everyone; as the work force is increased, new workers will learn to work at the rate. One must never initiate rates that will reduce wages *for anyone* or there will be no interest by the workers. The best time to start piecework is at the beginning of a new job.

4.19 ACCURACY OF SPOT CHECKS

A *spot check* is an observation in detail of a few workers for a short time, since it is too expensive to watch them all. All spot checks of production are approximate. A worker who happens to be moving from one location to another, or is talking or drinking coffee, is not producing at all. When everything is in order, a production rate several times the daily rate may be attained for short periods. Except for the trowel trades, who work with comparatively few delays, it is reasonable to halve observed spot production rates to estimate daily production.

Spot checks must, of course, be made when the workers affected are not aware that they are being timed. At first, this is not difficult; an occasional glance at a stopwatch on the wrist when talking with a foreman or looking at other work will not be noticed. Many items can be checked from a considerable distance. If you have the same crew for some time, however, they will eventually recognize what is happening, and you may have to delegate such work to others.

Although unions are traditionally against time checks of this type, your relations with the workers should be such that by the time they find out what you are doing, they will enjoy it. But this will make your observations useless—the workers are likely to speed up for a short time, to show what they can do!

4.20 USE OF OVERTIME

In some types of work, particularly pipelaying, there are many delays due to weather, pumping out trenches, and similar operations, which often make a long day economical. If the operation is going well at the end of the day, it may be economical to continue for several hours, even at time and a half. The foreman should check the production rate and continue work as long as this speed is over a certain rate, which must justify overtime. The workers, then, will do their best because they know that if the work falters, they will go home. In this way, for a period at least, the crew is actually working on piecework in a way about which even a union could hardly complain.

4.21 ESTIMATE OF COST TO COMPLETE

As a job nears completion, a more accurate estimate of work done can be made by listing each item to be done and the probable cost. You may be requested to make the labor and material cost part of this estimate, and the bookkeeper will furnish the amount still due subcontractors. This estimate is a basis for estimating the profit or loss on the job and also to judge the remaining cash requirement. You should be generous with the estimated amounts, as the last items are hard to estimate accurately. They are often left to the last because they are difficult. At this stage, your own salary will be a large part of the remaining cost, so be sure you allow enough time–not to do any work, but for the subtractors to finish up. You should not let your desire to finish the job affect your estimate of the time the subcontractors will actually take!

4.22 SUMMARY

Cost accounting is second only to reducing production cost in your duties as a superintendent. Learn your company's methods thoroughly–be painstaking and insistent until you know as much as there is to know about it. Some contractors have no cost accounting system at all; others have such an intricate and complicated system they are unable to summarize the information intelligently after they gather it. Most of them have been to both extremes and have evolved a workable but not overly ambitious scheme. Your own work should be as complete as your employer requires and, for your own benefit, as complete as available clerical help permits. If your own system isn't too good, use it anyway–you will learn about relative costs and will encourage your foremen to appreciate and measure production. And your jobs will show successive improvement in their cost accounting returns.

Scheduling

5

Dissatisfied worker to foreman: "All this planning is all right, but remember, Rome wasn't built in a day!"

Foreman: "Yes, but I wasn't running that job!"

For the most part, construction jobs are completed almost on a random guess of required overall time. Usually, the basis for the contractor's agreed time or the architect's specified time is the time that a similar job has taken before.

5.1 PURPOSE OF SCHEDULING

This being the case, why should the superintendent be concerned with planning the time for parts of a job, that is, with *scheduling?* Generally, the superintendent must do scheduling because no one else does. You are there, you are hiring the personnel, directing the work force, calling and pushing the subcontractors. You have no crystal ball, but you have a better command over the situation than anyone else. And regardless of how the estimated completion date was determined, you are stuck with it.

Who is harmed by late completion? The value of the job represents an investment, and all with money in the job are losing the income this investment could make elsewhere. This is usually much higher than mortgage interest—20 percent return on business investment (or 10 percent after taxes) is not uncommon. A $1,000,000 job may therefore be wasting $550 daily for delay near the end.

In some cases, buildings must be completed at certain times or they are useless for a considerable period. Department stores must be completed in September for the Christmas trade or they will not be profitable to open until March. Schools must be completed for the school year, resort hotels for the season. Usually, this cost of delay is paid by the owner; owner-built jobs are therefore inclined to be pushed faster than are contract jobs.

Some contracts require the contractor to pay *liquidated damages* for delays beyond a certain date; contractors do not like this arrangement, although the charges are practically always much less than the damages to the owner. Since contractors feel much more sure of their ability to control costs than to promise delivery at a specified time, they will bid considerably higher on jobs with high liquidated damages clauses.

The owner often chooses a contractor on the basis of his reputation for prompt completion rather than by competitive bidding. For this reason alone, it is important that completion dates be met. It also often works out that the properly planned job is less expensive than one built by day-to-day planning, because scheduling and cost reporting are closely tied together—a delay in the work is immediately recognizable as an increase in cost, and action is taken to correct both the delay and the cost increase.

5.2 HOW TO SCHEDULE

Scheduling is done by listing the various operations in a construction project by trade and *sequence,* estimating the number of workers to be used and the time required for each operation. Typically, work must be done in a certain order and at different times; the speed of the work will depend on the available number of workers of a required trade.

If proper attention is given to the operation that is holding up the work, other operations will usually keep up. Because each of the trades usually requires a different subcontractor, your job is largely persuading the subcontractors to get on the job and to furnish enough workers to do the job.

Because of unforeseen delays, a forecast of required time is always a rough estimate. You are limited by the number of personnel available, both yours and the subcontractors; and the subcontractors are unable to plan their work, as they are dependent on completion of work by other contractors. If all jobs in a locality were to be completed as planned, community-wide planning of the labor force would be necessary. This is impossible under present conditions in the industry, as there are so many small firms. Business agents for labor unions try to plan the available labor force, at least for larger jobs, by transfer of workers to and

from other areas. The business agents do not have any control over labor demand, however, and are plagued by periodic general shortages and surpluses of workers.

After a large job has begun but before workers of a certain trade are needed, a number of smaller jobs may start and draw off the available labor force. Most construction scheduling by college professors and engineers ignores this factor, largely because the future demand for labor is virtually impossible to forecast. There is enough shifting between jobs to make the labor supply always sensitive to unemployment or labor shortage in other areas or in other industries.

Because estimates are often in error, some consider scheduling of work impossible. This need not be so—most construction jobs can be planned with sufficient flexibility that scheduled completion dates can be made even if the number of available workers is unknown. Some operations at any particular time are critical—that is, they hold up other work. The concrete frame of a building is essential in the early stages, for example. But by careful planning the available workers can be used on these critical items, making it possible for more trades to work at the same time later.

By choosing the right places to start work, available workers can be used, even though you would have liked to work in a different order. If there is a shortage on one trade, or will be, arrange for those people to start work as soon as possible. In any case, allow extra time for unexpected delays.

The construction time for a project is the total of the necessary times for jobs which must follow each other, and a well-run job is consequently one on the line between trades following each other and overrunning each other. These jobs are made as small as possible; that is, you try to start each trade as closely on the heels of the preceding one as possible, but you allow for delay between trades. If you schedule the job to go faster than can be accomplished, you will be obliged to delay starting times for subcontractors. Once you have told a subcontractor you are behind and he must wait, you are unlikely to get him on schedule again. If you schedule the work to go more slowly than you are able to, the subcontractor is not obliged to move in early, and you must wait for him. It is preferable to be waiting for him: this gives him an option of starting work sooner and he knows he can plan on starting at least as early as the scheduled date.

5.3 TIME REQUIRED FOR EACH ITEM

After breaking the job down into trade items that must follow each other, you must estimate the time for each item. For heavy construction work and building work involving machinery, the work is planned for a machine

and a certain-size crew; this may be done before the job is bid. Items in building construction work are rarely planned in such detail, and the best information you can usually find is the estimated labor cost per item.

If you have to make schedule estimates, it will be one of two circumstances:

1. Cost or labor-hour estimates have not been made, so you must do a complete job.
2. You are given the estimate prepared by an estimator in terms of cost or labor-hours and are required to convert this estimate to calendar time required to complete the work.

5.4 COMPLETE ESTIMATES

If you must make a complete estimate, the first step is to make a cost estimate, just as is done for a bid estimate. Until recently, unit prices were used for estimating labor, bypassing labor-hours and hourly rates entirely. With the rapid increase in hourly rates, this may be done only for units used every year and adjusted for pay rates. This is acceptable for housing and many kinds of construction, but for firms doing a widely varying business, the readjustment of unit prices is more laborious than the use of labor-hours. Also, commercially available estimating guides are now almost entirely in man-hours. Determination of labor cost is outside the scope of this text.

5.5 CONVERTING LABOR ESTIMATES TO CREW TIME

From your own experience, you will learn the options you have in assigning workers to tasks. Since this account is primarily of building construction, we assume that the size of the work force determines the time to complete the work. In heavy construction, equipment rather than labor is critical, but the method is similar.

For the work as a whole, the labor-hours or worker-days for each trade are divided by the time available, to determine the required work force. Then by working through each part of the work for each trade, you can see if these completion dates will match the rest of the job, that is, that each trade will constantly have the work ahead completed for it and their own work will be completed on time for the following trade. A built-in lag must therefore exist between trades unless the work force is constantly changing. The superintendent can do a great deal to cut down

Figure 12 A strip shopping center.

these delays by planning; each plan poorly made is an aid to improve the next one.

As a sample, let's suppose you have a shopping center job of about 300,000-square-feet floor space, with the largest store a department store such as J. C. Penney (see Fig. 12). It is to be slab on grade, with masonry bearing walls and open-web steel joists for roof support. This is a simple structure, and most cities have at least one such project.

No amount of figuring can replace experience with the particular kind of job you are doing. To intelligently plan a job, you must have already built a similar one, or several. Scheduling is a trial-and-error process, where you plan what you are going to do, try it, and plan again with knowledge of your first mistakes.

Let's assume that you have a labor cost estimate for the job, broken down by items. If you haven't, you'll have to make your own, either using worker-hours or labor cost. Using the cost estimate, you plan the crew size necessary for the operation and figure out what your daily production must be to meet the estimate.

5.6 PLANNING THE FOOTINGS

Suppose that you are going to dig and pour continuous footings with a 9-person crew, and the estimated labor is 270 worker-days for 9000 linear feet.

$$\text{Required time} = \frac{270 \text{ labor-days}}{9 \text{ workers}} = 30 \text{ days}$$

$$\text{Daily production} = \frac{9000 \text{ linear feet}}{30 \text{ days}}$$

$$= 300 \text{ feet per day}$$

If you are not producing 300 feet of footings per day, you know you are behind both in time and cost.

5.7 SEQUENCE OF WORK

Next, you must decide what part of the project will take the longest time or will have to be delivered first. In this examplc, the department store is most important, for several reasons:

1. It must be completed much earlier than the others, as the tenant has a lot of fixture work to do before he opens.
2. It is in part two-story and will therefore take longer.
3. Other stores will not open until the department store is open, so it would be useless to complete other stores before the department store.
4. Since the department store has the largest amount of fixturing to do, more personnel to train, and more goods to buy, the owner will not schedule an opening unless he is sure the delivery date will be met. He will not be sure of this unless he can see substantial construction under way.
5. The department store has a larger proportion of subcontract work to structure, so it is possible to start more of the subcontractor's workers sooner if work is started here. This, in turn, means that all subcontract work on the job will be completed more rapidly.

You will notice that the superintendent may not know about some of these requirements, as they are either in the leases or in agreements between the owner and the contractor. You should therefore do your best to discover the conditions of delivery of the building before the job starts.

On the other hand, the Penney store has a steel frame; if structural steel were to be late, it might be of advantage to complete those stores that do not require steel. In this example, it is assumed that material delivery is not a problem anywhere on the job. The contractor must check material deliveries to see that requirements can be met in the manner that is most economical from a labor-cost standpoint. If material deliveries must be accepted at other than the desired time, the contractor must decide if you will reschedule work to meet material deliveries or will delay starting until the work can proceed continuously. In this example, we select the department store to be started first.

Should you start at more than one place at the same time? This depends on how soon the job is needed, when material will be delivered, and and the number of workers available. A crew that does the same work for a longer period usually is more productive, so we will try planning with one crew and see how much time is required. Where one crew is too slow, another crew is then planned. Of course, more than one crew in one trade may be necessary for some operations; if interior walls are block and exterior walls brick, there will be separate crews for each operation. But two crews would not be put on block work alone unless one crew is too slow.

Critical work is that which delays other work and the overall completion. Work that may be delayed without causing delay in overall completion is not critical. Obviously, any part of the work will become critical if delayed too long.

5.8 MASONRY

The 30 days required for footings may not be *critical* except for the department store. The footings need not go faster than the masonry, and the latter may be critical.

The masonry must start as soon as possible. If we have 300 feet of footings each day, we can start masonry material distribution the second day and laying walls the third. With 12-foot-high walls, 3600 square feet per day can be laid. At 200 square feet per bricklayer per day, 18 bricklayers are needed. If the wall is laid at one scaffold lift (4 feet) per day, the force may be put on at one-third per day, with a full force the third day. The total bricklaying time is then 30 days, plus 3 days to start, or 33 days.

However, more delay is necessary between the footings and masonry work for a number of reasons. It takes time to stock the wall and to erect scaffold, to level the ground which has been dug up for the footings, and some engineers will object to loading footings too soon after being poured. Some engineers also object to carrying up one wall too far in advance of adjacent walls because of differential loading of the footings. A work force cannot efficiently be hired overnight, and 30 days is such a short time that it would be difficult to get good workers or to get good production with so little work ahead. Particularly in the fall, personnel would hardly be hired before they would be looking for a longer job to last through the winter. On the other hand, if the job is being worked in the winter in a cold climate, there would be plenty of workers available, but it would be unprofitable to rush the job during the period of difficult weather and therefore of expensive work. These problems arise regardless of whether a subcontractor is used or if the workers are hired directly, but the subcontractor should be able to build up a force more promptly because he has other jobs from which to transfer personnel.

5.9 SLOWEST TRADE

At this point in the work—not 2 months later—we must decide which is the *slowest* trade on the job. Is there any reason to complete masonry in 30 days, or will the bricklayers just leave everyone else behind? So you must check the total labor force each subcontractor will have available—perhaps more than one subcontractor will be required for the same trade in some instances. You may be told that several subcontractors will not work on the same job, and some subcontractors will refuse to take only part of a job. However, there are few subcontractors who will refuse work merely because someone else is on the job. They may not like to work this way, because it means that extra work is competitively bid. A job similar to that in Fig. 12 may have three to six electrical contractors, two or three plumbers, and two plasterers. It may be useful to have one drywall and one plastering contractor on the job, if it is possible to change finishes to use the available labor force.

How do you identify the slow trade in advance? If hiring of workers is critical, the local business agent can usually inform you of the situation. He is likely to be optimistic about the labor supply, particularly in those areas where union clauses permit hiring of nonunion workers if the union is unable to supply workers within a certain time. But the number of workers may not be the important matter—you may be restricted to certain subcontractors, and they will refuse to increase their work force. They may not believe an increase in their work force is efficient, as new workers must be trained in the specialty.

In this case, let's assume that the *glazing trade* is critical. This presents three problems:

1. Glaziers are in short supply, limiting the production available.
2. Glaziers on store-front construction work most effectively by using a small force to set sash for several stores, and then borrowing workers from other jobs to set the glass. A glass truck is required for this, and the truck may be used elsewhere most of the time. Consequently, glaziers want several stores ready at a time, and prefer not to close up single stores. This means the preceding trades must be at least 1 week ahead of the glaziers, and trades that follow plate glass will be delayed for this week's working time.

3. Aluminum-store-front work is easily damaged and should not be exposed any longer than necessary. It is desirable that as many trades as possible precede the glazing, to prevent much damage to the aluminum work. With a limited number of glaziers, the order of work may have to be changed.

In this example, assume that the supply of glaziers is such that 80 working days (4 months) are required for the storefronts. Notice that this doesn't mean that the subcontractor will be on the job only 80 days; he may be making up door frames while other members of the work force are pouring footings, or still trimming out interior store-front doors as the keys are being turned over to the tenants. But work he does that is a *single operation,* and holds up other trades, is the part in which we are interested. This part requires 80 days. If the glazier is limited by total manpower, he may not be holding anybody up—the completion of the job will then be 80 days after he starts work with a full force. It may be possible that all remaining trades can be completed while he is working on the fronts. That is, other trades are scheduled to the time required by the glazier.

Now that you have decided the slowest trade to be 80 days, there is no reason to schedule the bricklayers to complete the work in 30 days. To be sure of getting out of the way, 70 days would be enough. Let us assume that we will use nine bricklayers for 60 days. Note, however, that if we start with more bricklayers and lay off later, we can start glazing (as well as all other trades) earlier, and reduce total time.

5.10 PLANNING STEEL ERECTION

Steel erection requires a crane and a minimum-size crew, usually five. The erectors work much faster than do the earlier trades. Suppose their work can be done in 20 working days with the crane (many minor items are done after the crane has left) but that the preceding trade takes 60. If the erectors are to work continuously, as they would like to do, they will not start until the bricklayers have been working 40 days. Such a lag will delay the entire job. The superintendent wants the erectors to make repeated trips to reduce this delay. The number of trips required should be stated when the subcontract (if subcontracted) is signed.

A large part of the wall structure must be ready before the steel erection starts; this causes further delays. In a wall-bearing job, more bearing walls can be made ready sooner if bearing walls are built first, end walls later. If the department store is critical in this example, the subcontractor makes a special trip to the job to erect steel for this area. Let us assume that the steel erector has agreed to four trips to the job; the trips should be as few as possible both because of the added expense to him—which will ultimately be passed on to you—and the fact that he cannot be relied on each time to get to the job promptly when called. If he is 2 days late on each occasion, changing from four trips to five may cause an additional 2 days' delay in job completion.

You can keep the subcontractor's workers on the job by a clause in his contract that he may not remove workers once he has started them on the job; it is much more difficult to force him to start workers on the job. If he doesn't have the personnel, no amount of pressure can help. But when his workers are on *your* job, it's someone else who has to worry about getting them, and increased pressure on the subcontractor by another contractor may make him finish your work sooner in order to start the next job. Even though the subcontract does not definitely say that the subcontractor cannot remove workers from the job, there are usually clauses regarding continuous work which can be used to keep them—you tell him that moving his personnel off your job constitutes an abandonment of the contract. It is for this reason that a special trip to the job to erect steel for the first store is justified. A saving of 3 or 4 days here means that the other subcontractors will start that many days sooner, giving them a chance to build up a labor force gradually.

For this job, let's assume that the steel erector will get to the job 2 days after you are ready for him. Of course, you have notified him a week before, but he will be skeptical about coming to the job until he knows you are ready. He therefore will probably make a tentative schedule for his crane but will actually check the job himself before ordering the crew to move. When you get better acquainted with him, and he with you, you may persuade him that he does not need to check the job himself. Let us assume that he will have his steel in place in 3 days, and remove the crane; and in 4 days more he will be through with the store and leave it ready for the roof deck.

5.11 POINT TO START

If the ground is frozen, it may not be possible to start work inside the building until the roof is on, and the full 7 days plus roof deck installation time must be provided for steel erection. But under normal conditions, work starts inside the building as soon as the bricklayers are out of the way. In this building, there will be interior footings, which can be built at the same time the exterior brick walls are under way. Fill is brought in and spread. With 3 days' crane work to erect steel, fill work and spreading may continue during steel erection operations, in different bays. But some operations do not proceed throughout the floor area, and the steel erection should be planned so as to first free those areas of the building where the greatest amount of labor of other trades is required.

This will usually be the toilet-room area, so you will try to get the plumber to work as soon as possible on the toilet-room rough-in. This area then becomes the critical part of the store—the plumber will have done 6 days' work at the time the steel erector is through with the building. But you must then have further work for the plumber, and must plan continued work in other stores. The toilet room will be important only until the floor slab is poured; afterward, you will have large showroom areas with finish and wiring that will require more time and labor than do the toilet-room areas.

If you have plumbing in both sides of the first store, you may ask him to complete one side only in the 6 days. You will then have half the floor area free and will be somewhat relieved of the pressure in getting areas ready for the plumber. In smaller stores, the plumber can work independently of steel erection—he can put in rough-in either before or after steel erection, as convenient.

The entire job can now be scheduled for the plumber. We have decided on a 60-day bricklaying schedule, and the plumbers may be in the first store as long as necessary to let the bricklayers get ahead. With stores 140 feet deep (a popular dimension), and enough bricklayers to lay 150 feet of wall per day, we can maintain about a store a day. With 30 to 40 feet average store width, or about 5000 square feet of floor, this production of concrete floor slab can be readily maintained. The steel erector and roofer will go much faster than this, and both will need to return to the job.

This means that the plumbing rough-in crew will do a store a day—certainly no more than two workers will be required. However, the larger stores will require more labor than this. The total plumbing labor must be planned for the job, with a lag between bricklayers and plumbers that will vary throughout the job. The plumbers may start with two or four workers, perhaps a week after the bricklayers; by the time plumbing rough-in of the first store is completed, the plumbers will be 2 weeks behind. They will then catch up as simpler stores are built and, you hope, finish a week behind the bricklayers.

The necessary force of plumbers is then such as to rough-in in 60 days. Although a detailed analysis of store-to-store operations can be made, it is generally a matter of judgment to estimate how closely the plumbers will crowd the bricklayers if too many plumbers are started at once.

5.12 EASY WORK FIRST

A very common error in planning is to do the easy work first. On engineered construction where payment is made by unit prices, the contractor will do the easy work first, for his costs are low, to get money as fast as possible and increase his available cash supply. There is nothing wrong with doing this if the engineer allows, but the contractor should plan his own costs and progress accordingly. This sequence of work usually does not delay overall completion time.

Another common error is more serious—to start the easy part of work in order to make a good showing, to the detriment of critical but less obvious work. A superintendent is often urged to do this by people who view construction as getting the most workers on the job rather than planning the necessary sequence of work.

5.13 HOW OFTEN SHOULD THE SUBCONTRACTOR CHANGE THE WORK FORCE?

Too many superintendents expect the subcontractors to appear and disappear at will. Their failure to achieve this expectation results in continued friction with the subcontractors. In preceding paragraphs it was assumed that the work force in each trade will be the same throughout the job; this is not necessary or even desirable in most cases. The subcontractor will require time to collect personnel from other work, especially if he must obtain a large work force. For each trade, it will be possible to complete the job on a given schedule with a certain number of workers, but if there are more available, it will be of advantage to use them. Since every person on the job must be moved on and off at least once, it need not be a hardship to a subcontractor to require that he build up a force, hold it as long as work is available, and then gradually reduce it. He is concerned, primarily, by a superintendent who requires workers on the job who must be pulled off for lack of work and then must be returned later. *You therefore should schedule all trades so that a worker once removed from the job need not be returned.*

5.14 SAMPLE OF TRADE WORK-FORCE SCHEDULE

Each trade should have a schedule of personnel required, the purpose or stage of the work in which they will be used, and the time of start and finish. Such a schedule is shown in Fig. 13. The emphasis on the Penney store is evident; with the exception of rough-in, work in this store is completed with workers who are an addition to the work force on the rest of the job. This diagram assumes a maximum work force of 10 people. Two people start rough-in and stay on this work; after 7 days (when the first slab is poured) another pair are brought in for topping off. At 20 days, two more workers start fixtures, and are joined by two more at

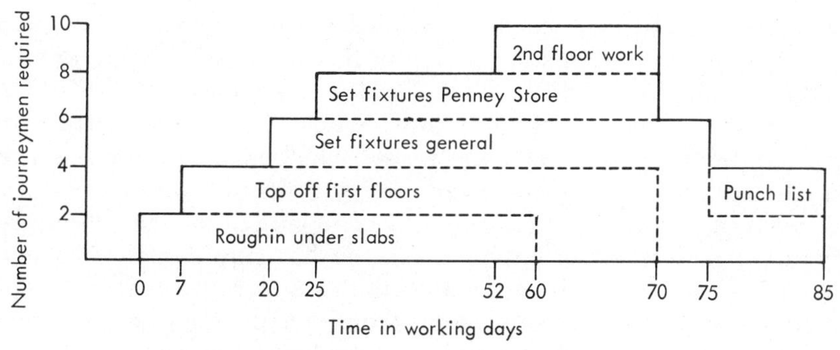

Figure 13 Plumber's manpower schedule.

25 days. Although the Penney store has priority, fixtures are started elsewhere; this is because the Penney store has tiled bathrooms and the other stores do not. It does not mean that the later store is less important, but if the additional personnel were not available to start the Penney fixtures, it would be necessary to rob workers from other stores for the Penney store.

5.15 INFORMATION FOR CONSTRUCTION SCHEDULES

Most superintendents don't like to make such a schedule for subcontractors, and contractors may actually forbid their superintendents to do so. The contractor may believe that specifying the number of workers a subcontractor must bring to the job will either make the contractor responsible to the subcontractor for delays if the job is not ready, or will relieve the subcontractor from responsibility for completing the job on time. That is, he believes the subcontractor will put the required number of people on the job but that they won't be enough, and then the subcontractor will say he has done as much as is required.

Very few contractors at present are in a position to accept responsibility for subcontractors' lost wages due to contractor's delay. The possible exception is foreign work, where it is necessary to recruit personnel and send them to the job before the job is actually ready for them. This schedule given a subcontractor, therefore, is not one the general contractor guarantees; it is what he intends to do. The subcontractor will acquire confidence in the superintendent's schedule only as it proves to be correct, or nearly so, on successive jobs. This confidence will be reflected in lower subcontract bid prices if a particular superintendent is to be used on the job, and on the subcontractor's willingness to pass up other jobs because he knows a superintendent's job will be ready when he says it will be.

Who really makes the schedule? Many superintendents refuse responsibility for a subcontractor's schedule, as this is the subcontractor's business. But the subcontractor cannot possibly make a schedule such as the preceding one, since he doesn't know when the various parts of the general contractor's work will be ready. The general's crew size determines the subcontractor's crew size, and yet the subcontractor doesn't know what the general is going to do. The superintendent alone can't make up such a schedule either, unless he has considerable experience in that particular trade. It is possible that a superintendent may know more about a subcontractor's labor requirements than the contractor does; for example, a superintendent who builds the same kind of project throughout the country could be expected to have a better idea of manpower requirements—even the plumber's—than would a plumber who did only one such project. Usually, the number of workers required must be determined by the plumber, but the amount of work done is determined by the superintendent.

For example, in this instance, the superintendent would ask the plumber how many workers it would take to do the first store and seven small ones in 14 days, or how many workers would be needed to do one store a day. These questions should not be asked offhand or without reference to the plans. Get the plumber to the desk, preferably with plans showing the piping layout, and make sure he has his mind on the matter. Be sure he isn't including excavation and time to wait for the inspector in his time estimate—this is time that overlaps from store to store. That is, one store is being piped while one is being excavated, and one is awaiting the inspector. The labor requirements depend on the *actual time spent working* on a store, not the number of days needed to complete one store.

Talk the matter over with the plumber foreman and with the subcontractor's engineer or field superintendent separately. If you get them together, the foreman will not give any ideas separately from his superior, and the engineer may be giving entirely impractical answers. Ask questions—of any kind. Your questions may make you unpopular, may show your ignorance, but they should have an important effect—they cause the subcontractor to think about the planning of the job. Subcontractors do not like to show their labor estimates, although many of them use very crude methods, such as a certain amount of money per fixture or labor cost as a percentage of the material cost. This means that although they may have a good idea of the total labor-hours required for the job, they have a poor idea of the time required for any particular part of it. If you ask for a written schedule, it will probably be made by the subcontractor's office man and will be designed to keep the subcontractor free of legal responsibility, rather than to get the job done. Subcontractors are often overcautious about what they sign, and for this reason it is often better to talk to the subcontractor's people and make out the schedule yourself.

In rough-in, for example, the labor should be nearly proportional to the number of joints. If you can determine the number of joints even approximately, you can compare the number of joints with the number of labor-hours and ask the reason for any large differences. If the plumber has not based his estimate on any real study of the plans, he will be obliged to do so to avoid, if not to answer, your questions! Since a superintendent is rarely familiar with details of trades other than his own, his ability to supervise them depends on his asking questions—to get information, and to see the subcontractor's reaction. If you ask a question that really isn't important or isn't one the subcontractor uses in his business, he'll say so, if he knows his trade. A person who doesn't know his trade, will attempt to cover up what he thinks he *should* know, and be slow to admit his ignorance. Generally, the more a person knows about a trade, the more he realizes the extent of his ignorance. Here again, your personal relations with the subcontractor or his foreman are most important—the foreman should never be embarrassed by his profession of ignorance.

The schedule you make out should not be considered rigid for any point. If the subcontractor cannot meet these requirements, he should expect that his contract will be canceled—not with bad tempers or arguments, but in the same spirit as when you offer a dime for a Coke at the corner drugstore, the clerk will insist on a quarter (or more!). It's just business. With the schedule, you still request workers when you need them, and if they are available in excess of the scheduled number, you may get more than you expected. If a subcontractor is willing to move on and off the job every 2 weeks, he may do so. You merely have minimum requirements as to what he should do, and he has a right to expect that you will not require certain things. But both sides do a great deal more than they *have* to do, and this covers errors in the schedule or the inability of one party to live up to his obligations at times.

5.16 ELECTRICAL WORK SCHEDULES

Although the electrician often delays completion of a project, he rarely delays the other trades. This is because there is not very much electrical labor that must be done before the finish trades can start to work, and some electrical work—particularly, pulling wires—may be done at nearly any stage of the project.

Electricians use a very detailed method of estimating, and members of NECA (National Electrical Contractors' Association) receive a book of labor-hours per piece or per foot for all types of material they install. Electrical contractors therefore would rarely use a less detailed method of estimating; they would either apply straight percentage changes to these figures or use a similar list of their own. Consequently, the electrician can readily figure the labor-hours required for any part of the work. This estimate is rarely checked in detail, so it still may be in error, but it is usually more satisfactory than that of other trades. The electrician will not have a labor cost by sections of the job, however, and it is quite a bit of work for him to estimate the work required for one phase of the work, such as, for example, installation of conduit under slabs. For small areas of electrical rough-in, an estimate by the foreman from the layout is usually as accurate as one made from the job estimate.

Where allowed by specifications and codes, thin-walled tubing may be used under or in concrete floor slabs, in lieu of the more commonly used rigid conduit. This cuts down on time and labor. When small pours are made on large floors, electrical conduit may be installed in each section being poured, allowing concrete trucks to pass through the building. Although electrical labor would appear to be greater when tubing is laid in small areas rather than run continuously across the building, it has been found that when electricians had a certain amount of work to perform to remain ahead of the concrete crew, by completing small areas each day, costs were lower.

Progress of the electrician may be gauged through most of the job by (1) how closely he works to the trade he follows in installation of fixtures and wall outlets, and (2) how closely he works behind his own piping crew with pulling wires. The first test is an obvious one; if he does not finish up the areas, the superintendent is promptly after him. The second check is often overlooked, so that the electrician bypasses wire pulling in order to complete more apparent work. In this way, he may get far behind without being noticed.

Power wiring of mechanical equipment is also often overlooked. Because there are comparatively few workers qualified to install power and control wiring, the superintendent should see that this work is begun as soon as possible and followed continuously. It is particularly difficult to find out how complete the wiring is unless the complete wiring diagram is studied by the superintendent.

5.17 SCHEDULING CONCRETE FLOOR WORK

Progress does not necessarily depend on a large work force; a force too large will run out of work or be laid off or be pulled out to another job. This is particularly true when pouring concrete slabs. Concrete finishing is often sublet because of the inability of the superintendent to schedule work so that finishers are continuously occupied. Since 2000 square feet or more a day is necessary to keep even two finishers at work, smaller jobs must be done by periodic pours. If the job is such that pours of this size or larger may be made each day, continuous operation should be considered. The savings in both time and cost may be considerable.

For the job we are using for an example, the operations preceding concrete pours are plumbing, electrical work, and grading. The comments regarding plumbing apply generally to electrical work. If pours of 5000 square feet daily are to be used—as previously mentioned, this will complete the floors in 60 working days, following the bricklayers—the work installed in the slabs can be completed in 1 or 2 days' work ahead of concrete pours. In the particular instance, you are delayed several days by the plumber. Assuming that the plumber requires 6 days to do his underfloor work (largely because you do not want him to put a large force on the job, as he could not keep working) and 1 day to clean up his area, you can pour the first floor area 7 days after the bricklayers have completed the first store. Note that it isn't really the bricklayers that keep you waiting this long, but the fact that the bricklayers must have the entire store completed before the steel erector will start.

The first floors to be poured will be under the area of plaster and partitions, and therefore after the first day's pour you can start working carpentry partitions.

5.18 USE OF PERSONNEL AT DIFFERENT TRADES

Open-shop work offers the opportunity to stabilize the work force and increase progress (which implies lower costs) by the transfer of personnel between trades, that is, using the same workers for different types of work. Jurisdictional agreements between unions and lack of training of workers in more than one trade prevent this on union work. The lack of personnel skilled in more than one trade make this transfer of workers difficult in open-shop work as well, but some workers are skilled in more than one trade.

On work with a high percentage of unskilled labor, particularly at isolated locations, production is low but workers can be readily transferred between trades. They are supervised by specialized journeymen or foremen in each trade.

5.19 SCHEDULING ROUGH CARPENTRY

You have the cost estimate for rough carpentry in the first store. Let us assume that the first store is 40,000 square feet, or eight pours. The carpentry must then take at least 8 days, if there is carpentry work on the last pour; if carpenters work faster, they will run out of work. There may be no carpentry work on the last pours—the last pour can be in the center area, free of partitions. However, if we plan the carpentry for 8 days, there will be no work at all for carpenters on the ninth; again, we must stretch out the work in order to work continuously. In this case, let's assume that we will take 12 days for the carpentry. With an estimate of 40 worker-days, the labor force is

$$\frac{40}{12} = 3.33 \text{ carpenters}$$

We will therefore use three, which will make the expected completion time slightly greater. However, the carpenters can start by cutting and partition makeup before floors are poured. Here, again, you have a choice between efficiency and early completion; are you going to start makeup of partitions on the earth, where cost is higher, or wait for the concrete slab? In such a circumstance, the superintendent for a general contractor will more often wait a few days. If he is working for an owner or where time is critical, he will start sooner at higher cost. If there are no material delays, the few dollars extra labor cost represents a day not just in completion of the partitions but of the entire project.

The superintendent usually cannot control material deliveries, and unfortunately must often lay off people or hire fewer, *anticipating* material shortages.

Although some superintendents prefer to give a concrete slab a day's drying time to harden before working on it, this is not necessary if reasonable care is used. Pickup trucks have been driven on day-old floors without damage. A floor hardens more quickly than is generally supposed—in one instance, small boys with small-tread bicycles were found riding back and forth on a concrete floor only 2 hours after finishers had left it, but the floor was undamaged.

In addition to the makeup of partitions, carpenters can also start furring exterior walls before slabs are poured. There would otherwise be several days delay in the plasterers starting after the slab is poured.

5.20 SCHEDULING LATHING AND PLASTERING

These trades are particularly difficult to fit in because they cover so much area in a day, occupy the area completely to the exclusion of other trades, and require several separate operations. In the example job, where the quantity of metal lathing and furring is large, it is necessary to make a complete schedule of work required and to be accomplished day by day. The best way to get data for this is from the lather's labor estimate directly; if you cannot do this, the lather's unit prices or unit prices obtained from *The Building Estimator's Reference Book* (Chicago: Frank R. Walker Company) may be used. Usually, lathers are readily able to remain ahead of plasterers; for example, on rocklath a lather can lath about twice as much area per day as a plasterer can complete. When wire lath, particularly furring, is used, the number of lathers required to stay ahead of the same number of plasterers may be very different from day to day. The lathers may then need to work several days ahead of the plasterers part of the time in order to complete areas of wire lath and metal furring before the plasterers catch up.

5.21 COMPLETION

Each trade is scheduled as described for the trades we have discussed above. This schedule is then shown as a graph or schedule or both. As you can see, most delays occur because of the delay between trades rather than the time actually required to do the work, so cutting down the lag, moving each trade closer behind the next, may help to finish the job faster than putting more personnel on the job. As the work force is increased, it may be necessary to increase the lag between trades, and if the increase is not well planned, the workers in one trade will run out of work.

5.22 DELAYS

The time schedule you figure out from labor requirements is a skeleton—you must make your own allowances for delays due to weather, inspection, changes, and subcontractor's failure to appear as promised. The number of working days must be adjusted to the calendar, taking into account holidays, weekends, and local events (in some areas, construction projects are practically abandoned at the beginning of deer hunting season!).

In short, the schedule is an aid to your trial-and-error system, and serves as a basis for an estimate based on experienced judgment. The method used for estimating the time for completion of each item of work will vary according to the type of work and the available data; only an indication can be given of some typical methods in this book. There is no standard, or best way; it is up to you to plan your job with methods suited to the job.

5.23 THE SCHEDULE DIAGRAM

We have described how a schedule is made; but how does it actually look, and how do you show it on paper? There are four ways to do this:

1. A list of dates.
2. A bar chart.
3. A labor graph by trades.
4. A CPM (critical-path method) chart.

5.24 SCHEDULE BY LIST OF DATES

If the job is small, particularly when many trades will move on and off the job, the dates for beginning and completion of each trade or each part of a trade is sufficient, and this can be made directly by judgment. In writing subcontracts, this list may be included in each subcontract.

Figure 14 is a simple example of such a list. The dates are in terms of working days, which is changed to calendar days after the starting day is set, by skipping nonworking days.

In Fig. 14, layout is to start on day 1 and will last 10 days, ending on day 11. On the next day, day 2, footings start; day 4, masonry; but non-bearing walls do not start until day 41. The "end"-day numbers are used to check the operation. Steel erection ends day 54, 10 days behind masonry on day 44.

If any other graphical method is used, you will need the number of days needed between operations (*sequence*) and length of operation; the starting and completion dates of each operation are then calculated and shown graphically.

When this list is made in more detail, often entered into a computer, it becomes a CPM (complete project management or critical-path method) chart.

Item	Begin	(Working days) Time required	End
Layout	1	10	11
Footings	2	30	31
Masonry (a) Bearing walls	4	40	44
(b) All other work	41	30	71
Steel erection (not continuous)	9	45	54
Plumbing (a) Roughin	9	60	69
(b) Top off	16	63	79
(c) Set fixtures	29	65	94
Concrete floors	15	60	75
Carpentry partitions	16	60	76
Lathing and plastering	23	60	83
Glazing	25	80	105
Painting	33	60	93

Figure 14 Work schedule.

For the job for which we have been working up estimated times, the schedule by listed events would look like Fig. 14. When the starting date is set, working days are converted to calendar days, making proper provision for expected delays. This schedule may be much more detailed than shown, depending on the importance of giving other firms exact dates. Such "exact" dates would of course be "safe" dates and include all allowances for possible delays.

The work schedule in Fig. 14 represents the minimum necessary for this job. If this analysis appears complicated to you, remember that you won't sit down and do this in an hour; it may take a week or 2 working as you have free time, to make a reliable schedule. Larger jobs have much more complicated plans, and *consultants* charge thousands of dollars to make them up after the hard part—the worker-hour and sequence data—has been done, usually by the contractor's estimator. In the illustration, no attempt is made to plan all trades; those which you believe to control the overall progress of the job are listed. Electrical work, outside work, and flooring, for example, are not shown at all. Judge which trades you believe will control and then anticipate other trades will meet this schedule. Some subcontracts require that the subcontractor work at such a rate as "not to delay other trades." This is obviously impossible; somebody has to wait on somebody, or everyone would be working in the same place at once.

Glazing, you will notice, was selected as the slowest trade, and on this sheet it still remains 10 days behind finish painting. The basic time for completion of an operation is 60 days, and this time has been used for plumbing rough-in, concrete floors, carpentry, lathing and plastering, and painting. Painting is started 10 days after lathing and plastering; if plaster is still too damp to paint at this point, all the work will be delayed.

There are a number of items of information lacking on this type of schedule:

1. It does not give the owner a schedule in money, which would tell him how much cash is required for payments at any particular time. The contractor, likewise, does not know his own cash requirement. In addition, some owners want a single percentage completion of the job stated. Such a number enables persons unfamiliar with construction to appear familiar with the progress of the work. The U.S. government, in particular, accumulates these figures for statistical purposes.
2. It is not adapted, in this form, to a quick check to find out what may be the effect of a change or omission. If concrete floors are delayed a week in the middle of the job, will the overall completion of the job be delayed? By checking through the illustrated schedule, it may be seen that this will delay the carpentry parti-

tions, which follow the floors 1 day. The lathing, which follows the floors 7 days, may or may not be delayed, depending on whether the partitions can be completed at that particular point in the work.

3. Since the lag between trades is shown only at the beginning and end of the trade's work, the schedule doesn't show where one trade may catch up with another, although the *average* time shows a lag. For example, there may be very little plumbing work in one store and a great deal in another; actual progress may depend on which store is done first.
4. The list doesn't provide a graphical, easily understandable method of illustration.

5.25 THE BAR CHART

The oldest and most common chart of job progress is the *bar chart,* Fig. 15. A horizontal bar graph shows beginning and completion of each trade, most of which are separate subcontracts.

Each subcontractor cooperates by submitting his own schedule. The cost distribution is usually made so that the general contractor's markup is in the early part of the work. (Figure 15 does not show the same data as does Fig. 14.)

The item *Site work,* for example, has a *10* at the bar corresponding to May 1; *40,* to June 1; and so on. This bar means that work is to be begun on site work during April, will be 10 percent complete May first, 40 percent complete June first, 90 percent complete July first, and completed during July. To determine if work is on schedule, the actual percentage of work done at the end of each month is compared with the scheduled percentage. The total work done at the end of each month is determined by adding the amounts for each item; this corresponds to the monthly invoice for payment.

You will probably not be required to make a bar chart, as this is done in the office before the job begins. Until a few years ago, this was the accepted method of showing the work schedule for the architect or engineer, and is still commonly used.

Since the percentage completion shown for each trade does not show *what* work is done, the bar chart can show good progress when critical parts of the work are incomplete. For this reason the chart is not sensitive to actual job conditions, although it does show value of work in place. Since the values plotted include both labor and material, some trades will have a very irregular amount of work in place although they have a constant work force. Plumbers, for example, put much more labor into the early part of the job when comparatively cheap piping is installed, as compared with later in the job when fixtures are installed. Mechanical

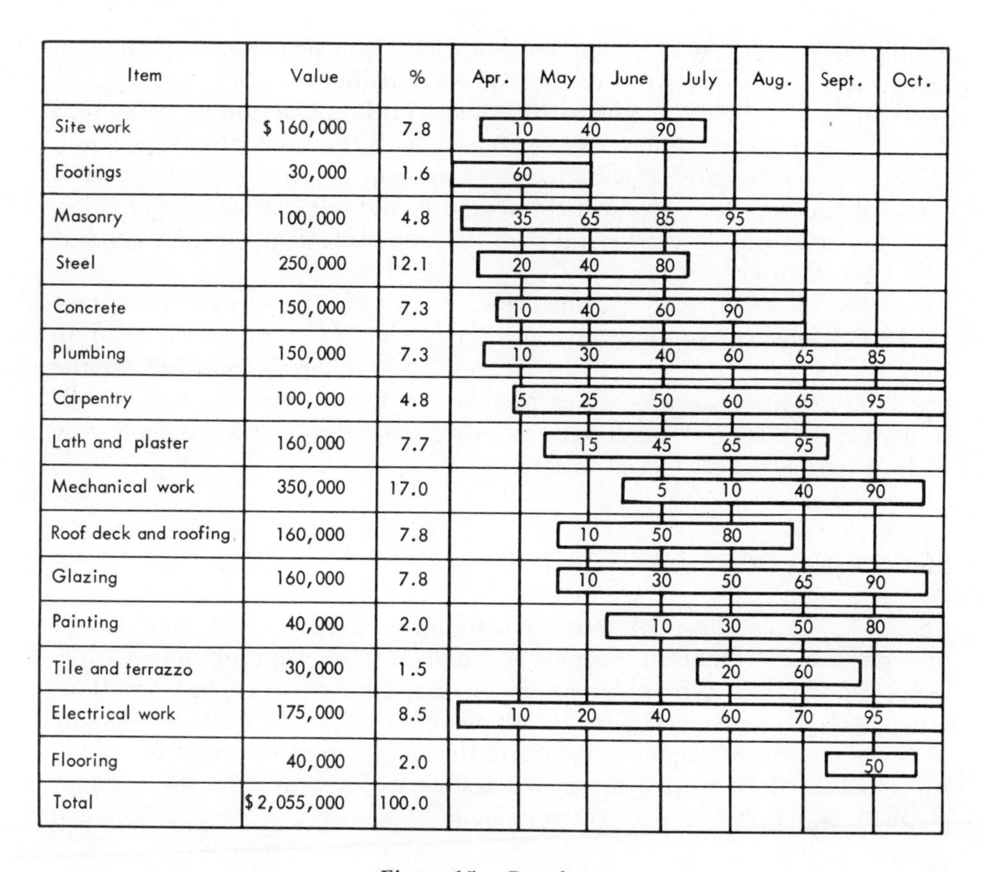

Item	Value	%
Site work	$ 160,000	7.8
Footings	30,000	1.6
Masonry	100,000	4.8
Steel	250,000	12.1
Concrete	150,000	7.3
Plumbing	150,000	7.3
Carpentry	100,000	4.8
Lath and plaster	160,000	7.7
Mechanical work	350,000	17.0
Roof deck and roofing	160,000	7.8
Glazing	160,000	7.8
Painting	40,000	2.0
Tile and terrazzo	30,000	1.5
Electrical work	175,000	8.5
Flooring	40,000	2.0
Total	$ 2,055,000	100.0

Figure 15 Bar chart.

contractors have large items of equipment which are installed with very little labor.

The bar graph could be a summary of a more detailed schedule; more commonly, it is merely an approximate estimate made in a few minutes by the contractor. If he wants to show good progress throughout the job, he will show large amounts for work done near the end of the job, although he expects to install the work much sooner. In this way, it appears that the job is ahead of schedule, and the contractor is freed from complaints by the architect during the job; by the time the bar chart shows that work is late, it is too late to do anything about it.

The bar graph is better than no schedule at all, and if based on a more detailed study of the job, will faithfully show the progress of the job. That is, if work is done in the best order and this order is not changed to make a good showing on the bar chart, the results of the bar chart are reliable. To make certain that this occurs, a more detailed schedule than the bar chart is necessary; an experienced superintendent carries this order in his mind and is fully aware when the bar chart is misleading. Anyone else, however, does not have this familiarity with the job.

You should never let yourself be lulled into satisfaction by being told you are ahead of schedule, unless you are sure the schedule itself is reliable. It is particularly difficult to get subcontractors to increase their forces if the bar chart (frequently posted in the job shack) shows the job to be on schedule. If you're not sure the chart is realistic, don't expose it to subcontractors.

In recent years, many writers have been pointing out the advantage of the CPM system as compared with the old bar chart, under the assumption that the bar chart was the method of scheduling followed by the contractor. This indicates a lack of practical knowledge; the bar chart is primarily to satisfy the architect or owner, but the contractor pays very little attention to it.

5.26 LABOR GRAPH BY TRADES

The chief disadvantage of most scheduling methods is that they do not provide the information directly needed by the superintendent—how many workers of each trade can be used, and how work-force variations affect the lag between trades. CPM planning does this only by making each operation very small indeed, and the lag inherent in computer operations makes the information needed slow in arriving. One more direct graph is shown in Fig. 16, which is made by plotting percentage for each

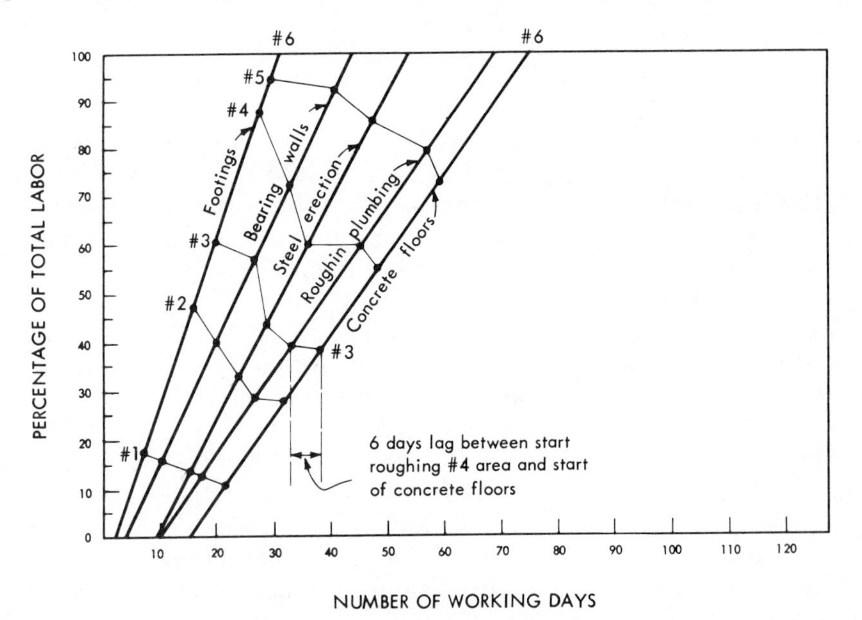

Figure 16 Labor graph by trades.

trade against the number of days worked. From this, the number of days lag between successive trades at any point of the work may be read. In the figure, footings are shown as starting on the second day and being completed on the thirty-first day, as in the work schedule in Fig. 14. If we are using the same-size labor force, the graph of footings will be a straight line. Now we go through each area (here we are using stores as areas) and find:

Store 1 requires 1530 lin. ft. of footings or 17% of total
Store 2 requires 2620 lin. ft. of footings or 29% of total

Total 1 and 2 = 46%

Store 3 requires 1250 lin. feet of footings or 14% of total

Total 1, 2, and 3 = 60%

If you figure the footage for each store and the cumulative amount at the time that store is completed, you can tabulate the results as follows:

Store Number	Lineal Feet	Percent	Total Percent
1	1530	17	17
2	2620	29	46
3	1250	14	60
4	2430	27	87
5	720	8	95
6	450	5	100
	9000		

From this list, you can plot on the footing line in Fig. 16 the point at which each store is completed, and from the horizontal scale you may read the anticipated date of completion of the footings for each store. Bearing walls are then plotted the same way, as are other operations following. For each trade, you will show the completion by a point; *1* marks the completion of store 1, etc., and the horizontal scale will show the number of working days each trade requires to complete that particular store. The horizontal difference between the completion point for two trades in the same store shows the lag between the trades at that particular point; from the plan of the store you can estimate the lag required to avoid conflicts between trades. This *lag* between trades indicates the number of working days the second trade requires to get out of the way of the first one. The lag can be quite small for some operations; in a large store, as many as six trades on eight to ten different operations may be working in one large room. The lathers installing channels for ceiling tile, carpenters installing the tile, and electricians installing light fixtures may work so closely behind each other that they appear to be a single crew.

This graph may be used to obtain schedule dates for subcontractors, and these dates incorporated in the subcontract or in a separate agreement. The percentages shown may serve also as a check on the progress of the work, and a point plotted on this graph from the actual progress and percent of labor spent shows both the progress of the job and the cost in relation to the estimate. If the point for completion of store 1 footings, for example, is to the left of the plotted point, it shows that progress is slow; if the point is *higher* than it should be, it shows that more of the allowed money has been spent than should have been and that the job is behind in terms of cost. A change in crew size changes the slope of the line, and a change in starting date of a trade shifts the entire line to the right or left.

From this graph, a detailed work schedule by dates may be made up. The work schedule may be made without the graph at all, but making the graph is only a little more work. This labor graph, as far as is known, is the writer's and you cannot therefore expect to find one unless you make it yourself. The other types of work schedules illustrated—a list of dates, bar graph, and CPM notation—are well known throughout the industry.

Each of the lines on the labor graph is actually a more detailed bar of the bar graph. If the percentages of labor required are shown on the bar graph rather than the percentages of work in place, the bar graph furnishes the data to plot the labor graph. The graph is made only to determine the lag between trades at any point. If all stores are identical—or if you have a housing job with identical houses or an office building with identical floors—the completion points for each house or floor will lie in a straight line, as is nearly true of the completion points for store 1 in the illustration.

In this and the preceding methods of scheduling, it is assumed that you can determine the important ("critical") operations at any one time by inspection. If the job is quite complicated, you may not be able to tell which operations should be pushed at any particular time; the *critical-path method* of analysis is intended to solve this problem.

5.27 THE CRITICAL-PATH METHOD

The list of dates shown on the Fig. 14 work schedule can be made very detailed and connected by lines to show a picture such as that of Fig. 17. For work so complicated that such a chart is necessary, the operations (or *activities*) are usually made by a computer program. The results are then available by a work schedule, or bar diagram, printed by the computer. The CPM chart is made initially in the office but becomes so large and unwieldy it is too large to understand and too difficult to keep current; the computer printout shows the needed information.

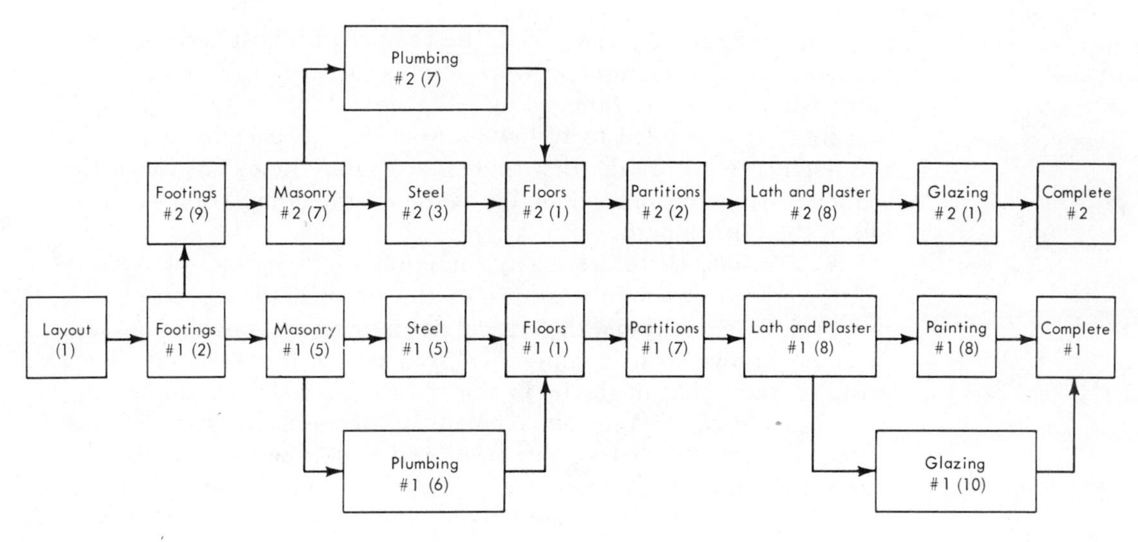

Figure 17 CPM diagram.

The work schedule (*computer printout*) can be made up as desired; the superintendent, for example, can ask for a list of operations *by starting dates,* so he knows day by day what must be done next. Also, he can get a forecast of personnel required, by trades. Most of the value of CPM planning to the superintendent depends on obtaining printouts of this sort.

In making the chart with this method, the various items of work that must follow each other are drawn in a line, and the total time for the job is figured in this way. Other series of items are drawn in another line and the total time compared. For example, in the work schedule in Fig. 14, completion of plumbing fixtures will be on day 94, painting on day 93, and glazing on day 105. None of these operations depend on the other, but all of them depend on some other work being done. If plumbing or painting were delayed, the job could still not be completed until day 105, after the glazing was completed; the glazing, or slowest work, is said to be on the *critical path.* This is illustrated in Fig. 17.

You will probably never be requested to make a CPM diagram, but you may contribute information for one and will often report completion of items on the diagram. There are several assumptions usually made to simplify these diagrams, and you will have trouble meeting schedules (or may improve on the schedules) if any of these assumptions are incorrect:

1. A job can be separated into work items, each of which must precede and follow another item. This requires the separation of the job into very small work items in many instances–much

smaller, for example, than would be required for cost accounting purposes. In this example, a work item is the work of a trade in a store which must be done before the next trade may begin. If one trade is scheduled to follow another into a space by 4 days, this 4 days' work by the first trade is a separate item. The longest sequence of operations, which limits the overall completion of the job, is the critical path.

As a result, there are a large number of items on the CPM chart, and various modifications to the original theory make it practical. For one thing, the critical path, or several possible ones, must be known as it is impractical to separate the work into all possible items that might be in the critical path. For example, if a wall is critical to the job, it might be necessary to start brick veneer on one end before the other end was completed; if there was ample time to complete it, the entire wall would be freed for bricklayers before they started. To consider all such alternatives, you would do much unnecessary planning.

2. The job can be planned without regard to whether or not the total number of people working at each trade is constant. You can assume you can get workers when you want them, and that the workers added are as good as the ones you already had; that is, that you can get the people you need, and that all workers are the same. The size of the work force is taken into account for each item but usually not for the job as a whole.
3. Some activities are not critical, as they require less time than is available. The number of days early that an activity can be completed is the *float time.* Delay of these activities will therefore not delay the work as a whole. Identification of these jobs allows personnel to be taken from them in order to complete the critical jobs; also, identification of the critical jobs may make overtime work on some of them profitable.

5.28 READING A CPM DIAGRAM

Figure 17 is a greatly simplified form of a CPM diagram for the first store of the job previously used as an example. On the left of the diagram, *Layout (1)* indicates that there is a 1-day layout job to be done before footings can be started. *Footings 1 (2)* indicates that the footings necessary for this store will take 2 days, following the layout, or at least the portion of the footings necessary for the masonry to begin will take this much time. *Steel 1 (5)* and *Plumbing 1 (6)* are shown as alternates in the chain of events, indicating that these items are to be done at the same time. Since steel is completed 1 day sooner, the steel erection is said to have a 1-day float time; that is, the steel may be delayed 1 day without delaying the next operation.

This diagram shows the total time for completion of store 1 to be

$$1 + 2 + 5 + 6 + 1 + 7 + 8 + 10 = 40 \text{ days}$$

Store 2 will be completed

$$1 + 2 + 9 + 7 + 3 + 1 + 2 + 8 + 1 = 34 \text{ days}$$

after the project begins. In this instance, the second store, because of the reduced labor requirements, will actually be completed before the first one.

5.29 PLANNING THE WORK FORCE

It is usually assumed that items which follow each other in the CPM diagram do so because of physical limitations, but this is not necessarily so. For example, in Fig. 17 the footings for store 2 are shown to follow those for store 1; this may be because the same crew is used. If two crews were used on footings, the footings for store 1 would follow the layout, thus starting—and finishing—2 days sooner. In this instance, this would complete store 2 two days sooner. However, this store is already 6 days ahead of store 1, so there is no purpose to this saving. Store 2 is not on the critical path, even if it must be completed at the same time as store 1.

If the entire job is needed and completion of some units or areas is not important, as in a single warehouse, school, or hospital, the operations completing the first areas are not important—they are ahead of other parts of the work and have considerable float time.

5.30 CRITICAL TRADES

When a CPM chart is first made, it is assumed that an unlimited number of workers are available when needed. Obviously, this is rarely true; the number of workers is limited both by the total available and by the rate at which they can be hired. If more workers are hired to complete the work as fast as possible, a point will occur when one trade will have more work to do than there are people to do it. If an item of work is on the critical path—that is, if it is delaying the other trades—workers of that trade will be diverted from other jobs in order to complete the critical work sooner. When another item, involving another trade, becomes critical, another trade will be delayed.

Consequently, no matter how the job is planned, it may suffer from a series of shortages of personnel in different trades. Scheduling the job requires an ability to foresee these shortages. Although it is possible to develop critical-path methods that will plan this work, in the present methods of making CPM charts, both manual and computer, this necessity for work-force planning is not stressed. The CPM chart may be checked for work-force requirements after being prepared, and this needs to be done. As originally drawn, the CPM chart also does not consider the necessity of keeping people working continuously. It is desirable to make necessary adjustments to the schedule by varying the time of completion of items with float time, but in some cases it may be necessary to extend the time of critical work by reducing the working force, to prevent losing the workers before some other job can pick them up.

An example of how the work force may appear at the time critical shortages of trades develop is shown by Fig. 18. At the beginning of the job, carpenters are needed for formwork. This work can be done more quickly, but to do so would require hiring carpenters more quickly than is efficient, and since there is less demand for carpenters later, it would mean that many people would have to be discharged. At the next stage, cement masons are needed to maintain a desired rate of concrete pour. It may be that there is a limited number of masons available, or for other reasons the rate of pour is limited. If additional cement masons are available, the superintendent may not be able to schedule concrete delivery or other items of the work, and the trade will still appear to be critical.

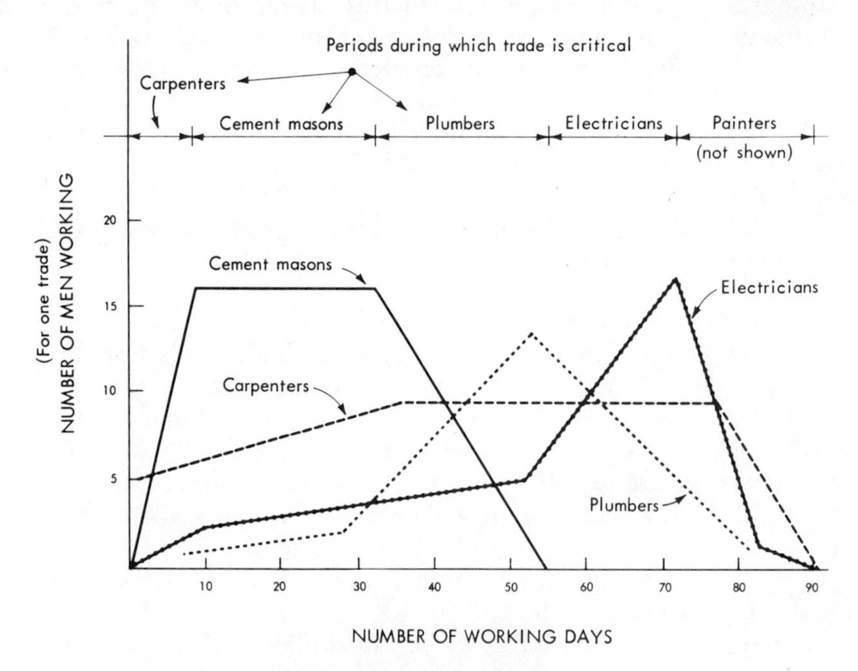

Figure 18 Critical trades.

In the illustration, plumbers are needed on the job faster than they are hired. By the time the needed plumbers are obtained, layoffs must be started; in the meantime, plumbers are the critical trade. *A trade is critical when the work force is either constant or increasing.* There should be no reason for a trade to be critical when the work force is falling, as workers would normally be retained as long as their trade is critical. Occasionally, a serious conflict between a subcontractor and the general contractor arises when the subcontractor determines to cut his costs by laying off less-productive workers, even though his trade is still critical.

In the example, electricians and painters are, respectively, the critical trades. All trades need not take their turn at becoming the critical trade; it is hoped that most subcontractors can do their work between other operations. In some instances, the same trade may be critical throughout the job. There will always be one trade which determines the speed of the job at any particular time; it is the superintendent's job to find that trade, preferably as soon as possible. It is not at all obvious which trade is critical at the time; the plumbers may work unnoticed for weeks, as they are not holding anyone up, only to find out that when everyone else is through, the plumbers have several weeks yet to go!

To repeat, the critical trade will be the trade working on the current item on the critical path, if the critical path is known. Merely making a diagram, however, does not ensure its accuracy, and there may be poorly planned or forgotten items that will delay the job. The superintendent should not depend on the CPM diagram without careful planning of his own.

A CPM diagram is a variation of a *flowchart.* A flowchart shows the physical movement of products through a series of processes, such as stone in a cutting mill or estimates and orders in a large office. The CPM diagram, on the other hand, shows a sequence of processes without regard to the physical location of the materials worked on. Flowcharts keep *space relationships* approximately correct; CPM charts emphasize *time sequence.* A shop flowchart may be the same as a CPM chart of the flow of a particular order through the shop, but it does not have to be.

Data for plotting the critical-trades chart can be obtained from computer programs, if work-force level is included in the program. If the critical-trades chart data from the computer are incorrect, the work-force size desired is fed back to the computer, which shows a new critical path. This should be done in the office before the job is started. A chart such as the one shown in Fig. 18 should be furnished to the superintendent.

If it is anticipated that workers will be scarce in some trades, you may prevent later delays in that trade by starting overtime early in the job. When not told otherwise, you assume that the job overhead is the only added cost due to delay—that is, if you do not have much to do, your wages are an added cost to the job due to delay. Although the superintendent's wages are usually considered a cost of extending the job, this is not necessarily true; the contractor may not have immediate work for the superintendent and therefore may prefer the job to last a little longer. Also, superintendents are usually very efficient workers when duties allow them to work with their tools, and produce low-labor-cost work as compared with carpenters, for example, even though superintendents are paid more.

If your firm is using a CPM system, you will very likely be required to report on the items on the CPM diagram. These reports are similar to the cost status reports described in an earlier chapter; you report the percent completion for each item at regular intervals. Since cost-breakdown items may be also activities on the CPM schedule, the same field report can show progress and costs.

The CPM chart or printout should give the same results as do the trade graph and cost report. There will be more items on the CPM chart, however, since some items of work do not require site labor and therefore will not appear on the labor-cost report. All subcontractors' work will require separate reports for progress, rather than cost. Many items, such as ordering material, submission of samples, preparation of drawings, and delivery of materials, are completed by others. These items are often on the CPM chart, which is an overall check by the contractor on job completion rather than merely a superintendent's check on the work the superintendent controls.

There are many variations on the use of the CPM diagram; it may be kept by the supervising engineer or by the superintendent, or it may be prepared at the beginning of the job for planning purposes only and never referred to again. The superintendent may not even know it exists.

In recent decades "CPM" has become a byword to describe processes and plans that have little relationship with the original critical-path method and which were often well established before CPM. For this reason, one's definition of what is meant by "CPM planning" shows his background and age. There is little importance to this, except that older workers dislike to hear scheduling and planning—even estimating—referred to as a minor aspect of the CPM chart. In recent textbooks and in magazine articles, CPM has become a household word for any process that is useful in construction planning. One may even refer to settling jurisdictional disputes by CPM methods; engineers have been guilty of similar terminology in the past, referring to "engineering reasoning" as any intelligent consideration of a problem, as though engineers had a monopoly on thinking.

5.31 THE WORK SCHEDULE

Previous diagrams have shown the number of working days required for the job. Once the starting date is established, the number of working days is estimated, allowing for all loss of time due to weather, delays in obtaining personnel, and delays in getting subcontractors on the job. In this book, it is assumed that delivery of materials has been taken care of by the contractor or engineer in the office.

The cost of delay may also affect the number of working days. Weather delays can be reduced or eliminated by protection of the work; this cost must be balanced against the cost of delays.

5.32 CHOOSING THE SCHEDULING METHOD

The various planning charts and lists described in this chapter are compared in Fig. 19. The type of report used will depend on what is important to your firm. A firm with little cash in comparison with its commitments, but with nearly all work sublet, may use the bar chart. A firm with many small jobs, or one that must complete portions of jobs, will be particularly interested in a work schedule. The labor-cost report is especially important for contractors with a comparatively large payroll.

These methods are samples; other methods of planning will usually include more than one of these systems in a single report, or may make a much more detailed reporting system of one of the methods shown. The

Name	Exclusive use	Disadvantage
Work schedule	Gives specific dates to subcontractors and trades.	Difficult to draw up for large jobs without graphical presentation.
Bar chart	Gives overall view of work in place and required investment.	Combines all work of one type without regard to details. Important delays may not show up.
Labor graph by trades	Shows time lag at any time between successive trades.	Assumes chain of events is known and substantially repeated. Lacks information on delivery of materials.
CPM	Shows sequence of specific items of work for planning job progress, from standpoint of work rather than from available labor.	Is quite complicated to adjust for availability of men, and if effective to show lag between trades, becomes "wallpaper size."
Labor cost report	Shows labor cost in comparison with estimate, and therefore is an estimate of profit.	Does not show progress of work, or any information on subcontractors or material deliveries.

Figure 19 Comparison of construction progress reports.

work order system previously described may be used to combine the labor-cost report, CPM report, and work schedule, and is the basis for other reports.

On the other hand, the CPM items or other large items of work may be entirely separated from the cost accounting system. This is especially useful where a large job has an accounting section for reporting costs and an engineering section for planning items of work. Some contractors use a *job plan,* a description of an item of work and how it is to be done, made by a separate engineering section and including a number of trades; for example, all work on the concrete floor of a large building may be covered by one job plan.

Job Administration

6

When one advances from journeyman to foreman, he must write reports on payroll, material deliveries, and accidents. As a superintendent, you have more reports, not only for your own foremen's work but also for the work of subcontractors. You must learn to recognize which items require your attention and which do not. It must be admitted that some very successful people have never learned to handle office routine. It is said that Napoleon allowed his mail to lie unopened for a month, on the principle that in that time most of the matters would have settled themselves and no action would be required. Many superintendents handle their correspondence in the same way. But you are not Napoleon, so don't act like him!

6.1 TYPES OF REPORTS

Some of the reports expected of you were described previously. These are the specific reports: whether something has or has not been done; that a certain quantity of material has arrived; that a certain amount of money must be paid to someone. In addition to these reports, you will be required to write letters and make requests in forms which are not prescribed—that is, in addition to approving or disapproving items, you must write about things that don't happen, and explain why (or ask why) they don't happen. Some examples follow.

6.2 DELAYS IN JOB REPORTS

Superintendents get behind in their paperwork for a number of reasons. They have tough decisions to make and don't like to think about a matter until it can no longer be avoided. For some, self-expression comes hard and writing is therefore painful. Some have found out that their letters bring criticism from someone—the home office, the clerk, a subcontractor, or the contractor—and try to take the line of least resistance by writing as little as possible.

These, of course, are never the reasons given. The superintendent says he has no time, or sometimes that if he handles the matter by telephone a letter won't be necessary. Or that he's forgotten it.

First, take the matter of tough decisions. You are a superintendent because you can make decisions—right or wrong. And there is no reason why, if you are a person of your word, you should not write anything you would say. Sometimes, it is much more diplomatic to write a little more carefully than you would speak—a curt verbal expression does not linger in the other person's mind, while the same expression in writing will be studied for some time. If a disagreeable decision is written, it is often a good idea to talk to the person involved and give him at least your sympathy. But a failure to put your decisions in writing will make subcontractors and others suspicious, and you will have even more letters to write. Confirmation of an order by writing, especially when a promise of action (such as *cancelling a contract*) is involved, is much more likely to get results than is a conversation.

Second, suppose you are not accustomed to writing and feel awkward about it. A secretary or dictating machine is especially useful; you just imagine the person you are addressing is in front of you and dictate as you would talk. You are not the only person who, when writing a letter, is carried away by the idea that his conversation is being frozen into deathless paper and ink. Write as you would talk; your letter is the answer to someone's question: "What do you mean to say?" The letter written as it is spoken appears a little peculiar—you look it over and wonder why it doesn't sound like the letters you receive. Your sentences will be short and choppy, the language colloquial, and there will be considerable repetition of words. Don't worry about this—you weren't hired to produce literature. The important question is: Is the meaning clear? If it is clear, you will in time learn to be more careful with words and sentences. There is no reason to try to imitate the style of others until you have read many, many letters; but as you concentrate on making yourself clear, your writing will improve. And you don't need to write at all if you have either a cheap standard cassette recorder or a more expensive pocket recorder. You may speak into these just as if you were reporting by telephone.

The third objection—that you are reporting, in most instances, unpleasant facts—cannot be avoided. The bearer of ill tidings is never loved, and you are doubly criticized when you are reporting your own mistakes. But by overcoming this tendency, you will in the long run make a better impression on your superior; he will eventually find out anyway, and it is better that he find out from you. Of course, there is a universal desire to withhold unpleasant information until you have some good news to go with it; if you do this, put your bad news in a file labeled "Pending Bad News" or something of the sort; at least it has been taken care of!

6.3 COMPLETING YOUR REPORTS

Failure to complete reports is for the most part your own fault, either because of the delays mentioned above or because of your attempt to do more than you can. Be sure the work in front of you isn't more than you can handle. There are three ways to cut your paperwork down to size:

1. Complete the work yourself in less detail, or by improving your methods. Sometimes you find yourself copying figures from one report to another. If the items you are putting on a report are from somewhere else, there should be an easier way to do it.
2. This easier way is usually to delegate work to a clerk. Any report or letter, or part of one, which requires merely that other people's work be summarized and sent to the home office does not require a superintendent. If you haven't a clerk who can handle the work, make it clear to your employer that the lack of a clerk is costing him money. A letter to the contractor marked "*Subject:* where you're wasting money" will always get prompt attention! Don't hesitate to tell the boss you're overloaded with paperwork, but be careful how you tell him. He doesn't care how hard you work, so don't complain. He is interested in money—so make every request one that will save him money. Don't say, "I have more work than I can handle, and . . . " but "You will save money with a clerk, so that I can properly supervise the labor you're paying for. . . ."
3. If you can't handle paperwork and can't get help, don't hesitate to pass work back to the home office. The people there are not so likely to have more work than they can handle, because they have the boss's ear and eye. He can *see* what they are doing; he doesn't know how busy you are. There are many ways to do this, depending on the organization and the people. Sometimes a reverse approach works—an office engineer afraid of losing his job may resent any of his work you do. You then give him more work

> by threatening to take it away—like leading a pig by the tail. You pull on the pig's tail in a direction opposite to the way you want to go—the pig, being contrary, will then go your way.

A certain period each day should be set aside for correspondence; early afternoon is a good time for this. It should be early enough in the day so that the work will be accomplished in spite of interruptions, which are bound to occur. If you can't get it done during the day, take as much time as necessary at night. Many people find they can complete letters and reports after working hours in a fraction of the time necessary during regular hours, because of the lack of interruption.

6.4 THE DAILY REPORT

In one form or another, all but the smallest jobs require a *daily report* from the superintendent to the contractor. These reports are to provide a historical record of the job and to notify the contractor when help is needed from the home office.

"Historical" sounds rather useless, and unfortunately many superintendents fail to properly record information that may be of use later. There are many happenings which at the time appear to be of little importance and are consequently forgotten by the next day. Some of these items become important. It is very difficult to prove that a subcontractor did not properly staff a job, or what work he did at any particular time. The daily report shows this. Sometimes structural failures occur, and there is a question as to the condition of the ground—both as to moisture content and to temperature—when the work was done. Engineers have a tendency to assume that the work was not delayed by weather unless you have a day-by-day report to the contrary. This information is entered on the daily report (Fig. 20).

The report *notifies the contractor* of trouble which he can correct, or where his influence is needed. Parts may be needed, extra work may be ordered by the architect on the job, union matters may be coming to a head, a jurisdictional decision may be made, material deliveries may be overdue, equipment furnished from the home office may be late in arriving or idle—the number of items is endless. Some contractors may require a separate report for certain items, and for important items you will write a complete essay report—as when a strike occurs or a wall collapses. The report on small jobs may be combined with the daily payroll. If requests are to be made of the home office on the daily report, write them in red crayon so they won't be overlooked.

Re: Daily Report - - Dullsville Shopping Center

To: John B. Quick

From: S. O. Battason Date 3-10

Weather: Max 20 Min 18 Sky CLEAR Ground 4" FROZEN

Work Force

Supv. and cler.	2	Plumbers	5	Glazers	–
Carpenters	15	Electricians	7	Painters	1
Laborers	7	Roofers	–	Flr. layers	–
Bricklayers	10	Lathers	2	Opr. engrs.	3
Tenders	10	Plasterers	5	Cem. finish.	4
Steel rod.	1	Steel erect.	8	Porc. enamel	–
Surveyors	2	Ceil. erect.	2	Terrazzo	–
Drywall	2	Gas lines	–	Air. cond.	2

Work Completed: Floor store #5, Plaster #2, Steel #10

Work Underway: Started correcting steel #7
Exterior masonry wall E. side Bldg. #1

Bulldozer: Grading floors

Ferguson: Front footings

Union: Op. Engr. on job - agreed to app. on conveyor.
Plumber's cont. expires Apr. 30 - No progress.

Tenants: Penney Inspector, Wallgreen A/C Engr. No complaints

Extra Work: #7 store dwg. rec'd today shows complete A/C installation

Rented Equipment: Grader on temporary roads.

Remarks: Bulldozer stripping treads - needs track work.
Brick color slightly off; we have two days' brick ahead.
Steel erection crew fired for getting drunk at lunch time and chasing bricklayers off scaffold with headache ball.
2 plumbers laid up with carbon monoxide poisoning from air compressor.

Sam

Figure 20 Daily report.

6.5 SAMPLE DAILY REPORT

Figure 20 illustrates the items of information that may be on a daily report.

Weather conditions may not only affect structural strength and the compaction of earth, but also may be used to justify time extensions. Weather is an *act of God* and therefore a justification for delay, but only if the bad weather encountered is worse than normal for a particular climate and time of year. This requires that the contractor know at the end of the job how many days of bad weather occurred.

The contractor's *Work force* is shown on the payroll but not as a summary. Also, he may be interested more in the number of workers available, in comparison with planned or past work force, than in the actual hours worked. The hours worked may be beyond your control because of weather, but if you haven't the required number of workers, you won't be able to regain lost time when weather improves. It is important to know the number of workers a subcontractor has on the job in order to justify cancellation of his contract for lack of progress. The number of workers, in itself, is not proof that he is behind, but it may be used in connection with repeated letters notifying him of specific shortages. Also, these records are valuable when planning the next job; merely by plotting the number of workers used on the last job, with known shortages which resulted, you can make a fairly accurate estimate of personnel required on the next one. One's memory is very poor in this respect; you may remember the number of people but have only a very vague recollection of the length of time they worked.

The *Work completed* may be of interest, depending on the type of contract. If a schedule has been made, based on specific portions of the work being completed at certain times, the completion of these items (as CPM items) may be reported on the daily report. For a shopping center, as illustrated, tenants often question the architect or contractor about the status of their stores; the daily reports can be used by the contractor to provide them with this information. The tenants may want to verify opening dates, or they may be considering a change in the store.

The items of *Work under way* shown are those which have some special meaning. Obviously, the report cannot mention all work under way. In the example, the steel correction is of importance because it is in an area previously completed and is therefore delaying further work in the area. The exterior masonry wall is of importance because of the low temperatures at which the work is being done, which might later either result in damage to that particular wall or may serve as proof that the method of laying brick in cold weather is adequate. (The writer has laid masonry at subfreezing temperatures without protection.)

The *Bulldozer* and *Ferguson* are contractor-owned machines permanently assigned to the job. The contractor is interested in knowing that these machines are properly used, in order to avoid idle machines on one job that could be rented or used on another.

The *Union* comments give the contractor prior notice of possible trouble spots. The number of strikes on a job is very small, but the possible disputes are large. The contractor needs assurance that you are settling union disputes as they arise, rather than ignoring them until a strike occurs. The first note—that an apprentice engineer has been hired—is of value to the contractor because he may have a blanket policy for all jobs, or may need information about the arrangements being made on your job to decide what to do elsewhere. The note of an impending expiration of plumber's contract implies that you are aware of the situation and are checking on what can be done about it. If a strike should occur, you have two obvious alternatives–to continue working with the same force, or to get new personnel.

The *Tenants* item applies only to this type of job–a shopping center in which the tenant is, in effect, the owner. Each store is built to a tenant's order, and he has much the same authority as if he were the owner on a contractor-built job. This particular job was built by an owner; in effect, there are two owners–the actual owner of the property, and the tenant who uses it. In most cases, the tenant is much more interested in the construction than is the actual owner of the property–who, after all, may not occupy it for 20 years. Tenants are interested both in quality of workmanship and in time of delivery.

The *Extra work* category is provided to give the contractor notice that costs will be higher, either because of changes by the tenants or architect, or because of errors in the work. By giving early notice to the contractor of such items, you may make it possible for him either to reduce the cost of the work or to collect from others. Extra work may also include subcontractors' correction of their own work; controversies often arise over the quality of work in place, and you should note if a subcontractor has begun correction of his work.

Rented equipment is here noted separately because of the high hourly cost incurred if equipment is idle or improperly used. A note on what is happening will reassure the contractor that the equipment is being properly used.

Remarks are items that may require attention from the contractor. In this instance, the bulldozer comment indicates that the machine will need shop work soon. The superintendent is not yet recommending it be done, as he needs the machine, but is merely getting the contractor used to the idea that there will be a large–and sudden–item of cost to be incurred later. If the contractor wants to check the equipment himself or have it done by someone else, he has an opportunity to do so; likewise, the contractor may want to transfer a machine from another job or buy a new machine. In effect, the superintendent is saying, "This machine will soon need a track overhaul. If you have a better idea, let me know."

The remark on the brick color indicates that the superintendent believes the range is satisfactory, or he would not accept the brick. Yet any decision regarding color depends on the opinion of the person accepting the building, and the contractor or architect should be aware of the variation and have an opportunity to see it. The superintendent is saying, "This brick looks all right to me, but there is a color variation–look at it yourself if you like." The note on brick shipment may be for the attention of the office engineer, or may be an observation on the effectiveness of the superintendent's own efforts, or both.

The note on the steel erection crew's escapade is a statement of a half day's delay on the part of the crew, but this is not the reason for its being included in the daily report—it is not that important. It does two things: prevents the contractor from being disturbed should he pick up a different version on his own, and is interesting enough to make him read the report.

All accidents on the job should be reported, unless they are very minor accidents on a large job. The carbon monoxide poisoning report is quite serious, as it could have been fatal; and in addition, negligence on the superintendent's part is shown. In this particular job, which was closed up for the winter, a gasoline-powered air compressor was allowed to run in a supermarket area where a number of people were working. Since enclosures were temporary and the job habitually operated with salamanders, no attention was given the compressor. The sickness of plumbers was not even recognized as carbon monoxide poisoning at the time. By implying he was negligent in this instance, the superintendent implies: "No one else recognized the danger of this situation, but I should have seen it."

The items on the sample report are real but did not occur on the same job or at the same time.

6.6 TYPES OF DAILY REPORTS

The use of an equipment operator's payroll report as a combined daily report, the *daily equipment report,* was shown in Chapter 4. For small jobs, two types of daily reports which combine the cost report and payroll report with progress and remarks are shown in the *daily construction report,* Fig. 21, and the *daily report,* Figs. 22 and 23. These are both standard pocket notebook sizes, sold by the Lefax Company. On such jobs, the material-received report may replace the receiving report, and instructions on the first blank given to the foreman may take the place of the work order. All information is then on a single sheet. Such orders are adapted to small additions or repair work. As jobs become smaller, the items to be shown are the same, but they may be combined into smaller and simpler form.

On larger jobs, a single report cannot show the detail that appears in Fig. 20. Besides, to whom is the report to be addressed? The illustrated report is for a job with less than 150 workers, having an office force con-

Daily Construction Report

TR.MK. REG. U.S. PAT. OFF.

Location ROSOFF Date 2/2

Weather FAIR

Equipment	Hrs.	Amt.	Labor	Hrs.	Amt.
B. DOZER	10		JOHNSON	8	
M. SAW	8		PERRY	8	
			JONES	8	
			SMITH	8	
Total			Total		

Material Delivered	Quantity	Material Delivered	Quantity
C. BRICK	12 M.		

Work Done: FINISHING GROUND AROUND BUILDINGS, CUTTING TILE FOR EXT. WALLS.

SPEC. FORM L-1369. PRINTED BY LEFAX, PHILA. 7, PA., U.S.A.

Figure 21 Daily construction report—A one-page report for small jobs which includes payroll and material deliveries (Lefax).

sisting of a clerk, and sending a report to the contractor who handles business matters for the job. If there are, for example, 500 people with a project manager and bookkeeper on the job, there is no purpose in sending a daily report to a supervisor of the project manager. A variety of reports may be used on such jobs, but they would generally be submitted to the project manager. These may be complete reports from each building superintendent, which would be similar to Fig. 20, or there may be centralized control with reports by trades and activities, or reports by other units than buildings. Each contractor designs for each job a reporting system that will tell the responsible person what is wrong and when he should take action.

DAILY REPORT

Date:
Job No:
County:
State:
TR.MK. REG. U.S. PAT. OFF.

Location ROSOFF

12-1-79

Name DAVID ROYER

Description of Work Being Done

CLEARING N. END
POURING FTGS. BLDG. #1

SPEC. FORM L-816. PRINTED BY LEFAX, PHILA. 7, PA. U.S.A.

Name	Occupation	Hrs.
ROYER, D.	SUPT.	8
JAMES, L. R.	ENGR.	10
SHAW, H.	CARP.	8
PRUGH, B.	OP. ENG.	8
TATE, H.	LAB.	8

Weather Report

A. M.	FAIR
P. M.	FAIR

Figure 22 Daily report (Lefax).

Total concrete poured 20 cu. yds.
Cement finished 1600 sq. yds.
Sacks of cement used READY - MIX
Lime used
Number of bricks laid 3000 @ $270. or 90 per M

Men	Trade	Hours	Wages	Remarks
1	Supt.	8		
1	Foreman	8		
1	Teamsters	8		
4	Bricklayers	32		
3	Carpenters	24		
2	Plumbers	16		
–	Plasterers			
–	Painters			
–	Roofers			
4	Laborers	32		
2	FIN.	20		
15	Totals			

Clearing Site		Woodwork	
Excavating–Grading		Roof Covering	
Reinforcing		Lath–Grounds	
Concrete	4	Plaster	
Forms	3	Plumbing	2
Structural steel		Steam	
Brickwork	6	Electrical Installation	
Stone–Terra Cotta		Marble	
Cutting Brick–Stone		Glazing	
Sifting sand			
Scaffolds			
Framing		Total	15

Special Instructions from Architect's Representative
SPACE TIES AT 15"
RATHER THAN 16" O.C.

12-1-79

Figure 23 Daily report (Lefax).

6.7 FILING ON THE JOB

Filing is looked upon with distaste by nearly all executives, particularly those who are accustomed to dealing directly with personnel. They don't see why it is of importance where a piece of paper is stored, and fervently hope they won't need anything that is over a week old.

Numbering systems are particularly disliked, because they don't tell you directly what is in the file. So files may have headings such as: *Superintendent* for letters signed by the superintendent; *Incoming* for letters received; *Outgoing* for letters sent (these two files alone could contain all correspondence); *Correspondence* for letters that arrive on cold mornings when the clerk doesn't want to walk to the incoming or outgoing files; and files by names of subcontractors. On one job there were twin files: *Heating and Ventilating* and *Ventilating and Heating*!

Numbering systems are necessary because there are a limited number of words that describe the categories of work on a construction job. The numbering system should be the same for cost items, correspondence, and payments; and the method of filing used by your company should normally be followed in the field office. If a numbering system is used, the letters you receive from the home office will have a file number; and letters you send to others will have a number on the copy sent to the home office so that their copy may be readily located when you want to talk to them about it by telephone.

What do you expect your filing system to do? The quick answer is "To find what I'm looking for." But it isn't that simple—just what are you looking for? If you know the date of each letter and the firm from which it was received, you may file by alphabetical order of firms, as does a bookkeeper. But this isn't what you normally look for: you want to know if a particular item has been ordered, and if so, from whom. You should receive copies of subcontracts and purchase orders from the office, and these may be kept in individual files if there are not too many and there is little or no other correspondence.

But just as soon as you must tell your clerk, "File this so you can find it again," you're in trouble. Suppose a letter came from the Ajax Steel Co., dated April 4, 1980, stating that they will ship a certain part of an order on a certain date. If you remember that the letter was from Ajax, he can find it again by a file of firms. But suppose you forget who is furnishing the material? You must then have some cross-indexing for firms supplying certain items. If you ask the clerk, "Have we a delivery date on widgets?" you will get a blank stare—he has no file of widgets.

The simplest job filing system is by specification sections for the particular job, and, if necessary, by specification paragraphs. If paragraph 10.12 is for *aluminum flagpoles,* you look for this number and name to find out if the flagpoles are ordered. If an item is in the specifications, you automatically have a file for it; items other than materials can usually be filed in the "general conditions" category. This means that your clerk is very likely unable to file correspondence, since he doesn't know all items by specification paragraph. You must designate the file for both incoming and outgoing correspondence. You then know, and the foremen know, where everything is filed. Only the filing clerk can't find anything, and he really doesn't need it, anyway! The masonry foreman, for example, knows that all correspondence and orders for his material will be in a certain file corresponding to the specifications—the carpentry foreman likewise.

Two copies of purchase orders are needed if job files are to be kept—that is, if correspondence about delivery and payment is to be handled at the job office. One copy is needed for the regular correspondence file and one for a *follow-up* or *tickler* file, which consists of all purchase orders filed in reverse order of delivery; that is, the next item to be delivered will be on top. All purchase orders should have an estimated delivery date; if the office does not put one on, do it yourself so you'll know how to file it to be on top when needed. Occasionally, you or the clerk go through the purchase orders and call the suppliers who should be delivering material for the next week or 10 days, to see if they are ready to make delivery and to order material shipped.

If you are approving invoices for payment on the job (which you will probably do only if your home office is in another city), you will need to keep delivery tickets during the month until invoices are received. This is easily done in an *accordion file*—a paper box with lettered slots in the top. The tickets are dropped into the slot corresponding to the name of the supplier; when the invoice arrives, you take out the delivery tickets that correspond to the invoice, check one against the other, and send them to the home office for payment. Often the contractor wants to know near the end of a month how much he will be billed for at the first of the following month. To estimate this amount, you take out all the delivery slips from the accordion file which show the material delivered during the month and figure the amount owed. Some items, particularly concrete, are ordinarily paid for at more frequent intervals, but the same procedure is followed.

6.8 FOLLOW-UP OF CORRESPONDENCE

If you have an inadequate filing system, or none at all, a natural result is a cluttered desk. If you have a letter on a matter that requires an answer or estimate from someone else before you can handle your part of it, you may not file the letter for fear of losing it. The result is a pile of *pending* papers on your desk, so if you must find the letter you want, you are forced to search through an entire week's correspondence.

It is common practice to file those letters which require checking after a period in a way that makes the letter or copy immediately available until the matter is settled. Some people use piles on their desk, putting less important letters or those referring to matters farther away at the bottom of the piles. Others use an index by dates, indicating when they must look at that letter again. In both cases, time is lost looking for letters that are "current" and therefore unfiled.

File all letters that require follow-up, and note on a desk calendar the appropriate day that you must check this matter. You then have the items available in the file on call. If you have an adequate numbering system, you need only a note of the file number on your calendar, so that you are reminded to look in that file on that date. Notations of telephone calls received should be made on slips of paper in the file, so that you can readily locate them also. A filing system made under the assumption that the letters in the file are unavailable is no filing system at all; you should be able to reach your secretary by telephone, to locate any letter in a few seconds.

6.9 THE THREE-PERSON OFFICE

A supervisor makes best use of his time in a three-person office. He rarely can do his own typing; but, in any case, he can save time by hiring a clerk to do not only typing, but routine reports, payrolls, and cost reports as well.

A stenographer, therefore, is necessary on any job where the home office is not readily available by telephone. On a small isolated job, it is less expensive to handle stenographic work at the home office, using the telephone and an office recorder. The tapes from the recorder can be mailed to the office and transcribed into writing there.

On many larger jobs, considerable time can be saved if a person who can give technical information to foremen and handle routine purchases is always available in the office. On some jobs, he may handle the routine of purchasing and negotiating subcontracts. This person may be called the *project engineer, office engineer,* or *assistant superintendent;* his function is to take care of the matters which, although routine (that is, they do not require experienced judgment), do require more technical knowledge than would be expected of a clerk. If the superintendent has estimating experience, the office engineer may also do estimating, so that the job office really becomes a district office.

6.10 SUMMARY

Paperwork is essential to get the job done and to satisfy the office that you are getting it done. By treating the handling of reports and letters as another item of the job—as far as you are concerned, another trade—you can set up a way to keep correspondence up to date with the least amount of work, and particularly with the least amount of your time. Don't expect your work to vanish if you don't think about it!

Labor Unions and Business Agents

"I hear the workers are striking."

"What for?"

"Shorter hours."

"Good luck to them. I always did think 60 minutes was too long for an hour!"

To make money for your employer, you must do two things: learn what your personnel have to do, and persuade them to do it. In this book, it is assumed that you know what is to be done; but how do you persuade people to do it?

7.1 PUBLIC ATTITUDES TOWARD UNIONS

If you have had little experience with labor unions, you may be inclined to view them with fear and suspicion. You have heard stories and read newspaper accounts about unions, and have possibly formed a distrustful attitude. This is to be expected; you are a superintendent because you are primarily interested in production; production means people must do what you tell them to do. How can they do that if they do what the union tells them to do?

The matter is not simple. Union organization is a world of its own, not subject to generalities and rigid rules. You like consistencies, clear-cut principles, rules that are universally followed by the parties who accept them. Unions just aren't that way. In some respects a union is a business, selling the labor of its members; but yet, unlike other businesses, it can't sell a uniform product or guarantee delivery. Other businesses can cut down their product or increase it; a union has a fairly fixed number of members.

Some areas of the education establishment are slanted against union organizations. Teachers often have had little contact with labor unions; at best, their attitude reflects ignorance, and often they are antagonistic. Newspapers often give labor unions a bad press for a number of reasons. As employers, publishers are likely to identify their interests with other employers. Reporters and even editors are not well paid, in comparison with the skill required, and may be jealous of craftsmen. Most important, the events that are "news"—matters that will sell newspapers—are events such as picketing and strikes, which are triggered by the union. Who is interested in a picture of an employer who departs for Florida rather than negotiate with employees? Pictures of picket lines and idle jobs, however, can be shown as "events."

The details of working conditions never come to public attention. A surprising part of the American public believes that construction workers are paid by the month, or at least by the week. When the railroads were running a publicity campaign against "featherbedding" by engineers who received a day's pay for a few hours' work, no mention was made of the fact that these men were required to travel to and from their job, sometimes several hours, on their own time, and often to lay over for several hours without pay.

This is not to imply that the labor union is or is not right in any particular instance; it is merely to point out that reading of the daily newspapers is an inadequate way to get adequate information about labor disputes.

7.2 LABOR UNION STEREOTYPES

The acts of individuals in the labor unions have often been publicized as being indicative of unions as a whole. Some superintendents with personal experience in towns where the labor unions are particularly restrictive or who have had jobs attracting the attention of unions tend to become bitter, and attach their resentments to other local agents. Some of the popular assumptions about unions are discussed below.

7.3 UNIONS RESTRICT PRODUCTION

The reason unions restrict production is, presumably, to make more work for their members. There is no question that this is done in certain areas. But it is by no means a characteristic only of labor unions; individual workers slow down considerably when the end of the job is in sight. It is well known that businesses with anything like a monopoly position do the same.

On the other hand, some unions, particularly the lathers and plasterers, realize their trade is threatened by other types of construction and do all they can to increase production. Apprentices may be required to do a minimum daily quota before becoming journeymen. Studies made by universities and by the government indicate that there is little or no difference between production on union and open-shop jobs, and in some cases the union jobs appear to be the more efficient.

7.4 UNION DEMANDS ARE UNCOMPROMISING

At various times, important jobs are held up because of very minor items, and the union finds itself in a ridiculous position. For example, there is a story of the mechanic who fell asleep on the job; he was discovered still asleep after working hours, and was fired. On appeal to the union, he was reinstated, and given overtime pay for sleeping after working hours! It is possible that such an event could occur; it is also possible that the worker had been told to go to an out-of-the-way corner and to wait for the foreman, who forgot about him. Because of such reported incidents, the attitude that unions have but one line—demands which they do not compromise—has become popular. As a matter of fact, compromises are made daily. A trade may claim certain work that could readily be done by laborers, where the claim of the skilled trade is doubtful. To keep the cost down to that of laborers, the business agent may offer to staff the work with apprentices and overlook the rules regarding percentage of apprentices.

7.5 UNIONS ARE RECENTLY FORMED

In many areas, unions are becoming more powerful; in others, they are losing ground. This has been happening for nearly a century. Before 1900, Frank B. Gilbreth's instructions to his superintendents included an order that, "Union laborers are to be given preference at all times, but no nonsense is to be taken from them." Employers have made efforts to prevent union work entirely by obtaining the cooperation of building material dealers who would refuse to supply union jobs.

The power of labor unions is often laid to a particular government, depending on the age of the person giving the opinion. Those contractors who started in business during the depression period of 1930-1945, when unions were weak, may regard Franklin D. Roosevelt as the man

who started unions. Those who started later will choose some other administration. The fact is that labor unions have been with us quite a long time—the first recorded strike was on construction in ancient Egypt. It is therefore not reasonable to assume that the basic problems will change, although the laws and ground rules change from time to time. The recent substitution of litigation for violence is certainly an improvement. By providing legal rights to labor unions, they have largely been persuaded to obey the laws.

7.6 UNIONS CREATE TRADE RULES

Whatever happens in the construction business, particularly with regard to labor, is often laid to union influence. Labor unions represent their members—and usually their members are about the same as everyone else. Trade or working rules adopted and usually enforced by labor unions specify a wide set of conditions. Indirectly, output is set by the number of workers required on crews, by requirements that workers be employed for certain work, and by the methods of work that must be followed. These rules have the effect of freezing work methods, but they were not originally dictated by unions. The difficulty arises from the fact that unions are usually the most conservative people in the construction business, and resist changes of any sort. Union leaders are interested in reducing friction: satisfied and working members pay dues and keep their officers; dissatisfied, striking members are not likely to become attached to the leaders who allowed the situation to develop. When the union takes action, it is usually because it is demanded by a significant part of the membership—even the most flagrant violation of working rules is allowed by the business agent if the members are not interested. Jurisdictional rules are intended to avoid disputes, not to cause them, and union leaders themselves are the people most anxious to settle disputes. Unions follow trade customs and the desires of their members; union leaders do little to change either.

7.7 UNION MEMBERS ALL HAVE THE SAME ATTITUDE

To the outsider, a labor union is made up of the same kind of people, all supporting the union actively, or controlled by it, depending on the point of view. But union members are as independent as any other workers—probably more so. One individual may be a member of one union one day and another union the next—usually forbidden by union rules. He may be

a union worker in Michigan during the summer, and an open-shop worker in Florida in the winter. He may work as union during the week and open-shop on weekends. Left alone, he has no regard for jurisdiction whatever; a bricklayer foreman may operate a bulldozer.

Young workers may want long hours—and more money for a growing family. Older workers may want a short week and more leisure. Consequently, a dispute arises over whether overtime should be demanded. Some people are better workers and better producers than others, and are much less dependent on the union; they will not enthusiastically support any union regulations designed to "help the weaker brother," which will ultimately be at their expense.

7.8 UNIONS ARE "ALL THE SAME"

Of course, unions have different *charters* and rules. Beyond that they have widely different attitudes toward the employer. Unions agree there should be a minimum wage (as high as possible, of course) but treat all other matters, such as selecting a foreman, jurisdiction, hiring, and discharge, differently.

By the nature of their work, some trades, such as cement masons, are hired by the day. If they do not support the union, they may therefore be out of work the following day, so the business agent has a powerful lever. If a union consists of workers employed by established contractors with a steady volume, the union will not be greatly concerned with measures to assure that all their people are working—they have already attained this. On the other hand, a union in an area with large and greatly fluctuating construction contracts is constantly concerned with employment of persons attracted to the area when work was more plentiful.

Electricians, with a long apprenticeship program and city licensing of journeymen to qualify their members, do not have the problem of unemployment that carpenters regularly have. Consequently, it is to be expected that seniority will be a more popular issue with the electricians, who are nearly always employed, than with the carpenters, who do not want to see themselves shut out of work by others. Seniority rules are supposedly a goal of all unions. However, it is not an overall gain for the workers, who, in construction, prefer to take their chances on unemployment in other ways.

7.9 CONSTRUCTION JOBS ARE EITHER UNION OR OPEN-SHOP

A construction job is quite often performed by a mixture of union and nonunion contractors. The general contractor may be open-shop, yet he often employs a small proportion of the total number of men. The

general may work one job open-shop and another union shop, either with two company names or with a single organization. Where two companies are used, one for union and one for open-shop work, no one is fooled; there is usually no attempt to cover up the common ownership of the companies.

A contractor may work under one name with a union contract, and under another as a union shop but signing no contracts. A union contract is not usually necessary to employ union members in the construction industry; in fact, the contract may be a liability.

7.10 THE SUPERINTENDENT AND THE UNION

The conceptions we have discussed are often not only assumed by the popular press, but may even be accepted as the basis of labor laws. As a superintendent, you will probably have little or nothing to do with labor regulations in a formal sense—that is, regarding wages, pensions, contracts, or general strikes to increase such benefits. But you have a great many negotiations on a day-by-day basis, where the feelings of each individual are more important than is the formal contract.

Remember, the union and the workers are constantly concerned with one idea: "Will this help or hurt me?" You can use the age-old principle of making a donkey go—put a carrot in front of his nose and poke a stick at the other end. You are equipped with various carrots and sticks, and so is the union business agent. Hopefully, you try to get him to cooperate so the job goes the way you both want. Where it appears that you cannot agree, you have the carrot—more money—and he has the stick—union fine or suspension. Your job is to get the workers to cooperate with you without bringing either them or you to an open break with the union.

The foregoing defense of labor unions has been intended to prevent you from allowing an emotional outlook to color your relations with the union. A great deal of friction between superintendents and union representatives has nothing to do with the matter under dispute, but occurs because of the determination of one party or another to win a point. You should be well acquainted with the business agent—well enough so that he will stretch a point for you on occasion, as you sometimes do for him.

7.11 DEAL WITH THE INDIVIDUALS, NOT WITH THE UNION

The union has been here represented as union officials, not as the body of your workers. The superintendent deals with both union officials and workers as individuals. The superintendent is acquainted with the individual workers—or he has foremen who are. You are not interested in obtaining "typical" workers but superior ones; to the extent that you are successful, the union goals do not represent the goals of the individuals. If a person known to be a good worker has a grievance, he has no reason to complain to the union; he merely moves to another job. His grievance is thus settled, and he will not be called a "trouble-maker."

When jobs are scarce, workers will call on the union for help more readily, and the business agent will be more anxious to place more workers on the job.

You may or may not be on good terms with the business agent; but there is no doubt that you must have satisfied workers on the job. Construction tradesmen don't expect a great deal, but you have to be as considerate of them at least as the average superintendent, and preferably more considerate. Union trouble is usually trouble with your own workers which has not been properly handled and can be stopped at the foreman level. Of course, this is by no means always true (nor is anything else in this business); you may have a determined effort by the business agent to take over the job completely, or action by the business agent that results from complaints by unemployed people who have never worked for you.

7.12 DUTIES OF THE UNION REPRESENTATIVE

The salaried official of a union local (some locals have several) is the business representative or business agent, commonly called the BA. He is usually required to have been a union member of that local for a number of years and is elected by vote of the membership. He may hold office one or more years, but there is a limitation under federal law on his maximum term of office. He is usually paid the earnings of a journeyman in that trade; his pay therefore goes up when his members get a raise. He is nominally ranked below the president of the union, but his authority rests directly on a popular vote of the membership. This authority will vary with the trade and with the local; generally, he has sole authority to call a strike or walkout (although he may have to obtain authority of the membership to do this), or to picket, and he negotiates with other unions on jurisdictional matters. He acts as employment agent and is the person contacted to send workers to a job. According to recent laws, an employer may not get employees from an exclusively union source, but the business agent may operate a hiring hall that also places nonunion employees. It appears that this rule is easily circumvented; one way is to give certain preferences to people who have lived in the area or who have been employed by certain contractors; these, of course, are union members. The title of the local union representative varies, and in some trades he is no longer called a business agent.

Business agents are politicians, not dictators; their job is to remain in office by pleasing their members. The most important thing a business agent can do for his members is to find them jobs, and the worst thing he can do is to take an action that will result in fewer jobs for voting members. Consequently, he can be expected to oppose not only entry of workers into the trade or union but also the entry of union members from other jurisdictions. In some areas, blacks are still opposed, and there appears to be little difference between areas of the country, in this regard.

It is of little avail to argue with a business agent regarding what he should do, if doing it would get him voted out of office. On the other hand, if you can help him put a few of his members to work, he may be willing to forego some benefits to which he is entitled. Concessions made to the union representative are not necessarily permanent; for example, an oiler was placed with a small, previously nonunion contractor on a half-yard backhoe rig. Within a week, the oiler by his own choice had abandoned his nominal job and was operating a bulldozer for backfilling. In order to keep two men on the backhoe, the BA would have had to check it himself. However, the person in question was working, so everyone was happy.

In another instance, two plumbers were hired to run water mains. This is hard, dirty work. Few plumbers care for it when other work is available. So, one plumber stayed as hooker-on on the bank, allowing the laborers to the pipe work, and the other plumber found another job. There are countless instances where the union wants jurisdiction over certain work, but its members will do the work only when none other is available. As the BA is the employment agent, he must *send out* the member who has been *on the bench* (without work) longest, or have a good excuse for sending another. You should specify the skill you need; few BA's will send an unskilled person to you if they can possibly avoid it. Few BA's are concerned if a member is discharged, either; they try to find jobs for their people, not to keep them on. The worker has to do this himself, and if a person is fired, the BA has an opportunity to find another member a job. The BA is therefore rarely concerned that someone has been fired, unless it is connected with union activity. When the BA defends a person as a good worker, the BA has two things in mind: to tell the worker the reason for layoff, and to reassure himself that the layoff was not for union activity.

Consequently, the timing of layoffs is important. You have a nearly unlimited right to discharge people, but not for union activity (unless it interferes with work), either by custom or law. (In union matters, custom is usually more important than law for the superintendent.) People should be discharged by their own foreman. You should not discharge someone when it can be interpreted as being due to union activity, and must give lack of work or inefficiency as the reason for discharge, if you state one at all. Much union trouble is due to a superintendent's talking too much or acting hastily.

7.13 HIGH COSTS CAUSED BY UNION OPERATION

There are several basic union principles that interfere with low-cost production. These will most often be the cause of your troubles with the BA, although the principles themselves are not mentioned. These are:

1. Wages.
2. Partiality between contractors.
3. Trade jurisdiction.
4. Hiring.

An explanation of these follows.

7.14 WAGES

Since setting of the wage scale is not done by the job superintendent, there is no conflict on this score. In some cases trades receive different scales for different work, and the superintendent must make a determination that the BA will not like. For example, a carpenter may receive more money for working on a high scaffold—but is the higher rate to apply to just the hours he works there, or to the entire day?

The superintendent has no incentive to pay lower wages for regular time. Virtually every incentive pay scheme in this country—including those to which labor unions object violently—requires higher wages. Efficient production requires cooperation and the most skilled people; such people are rarely available at lower than standard wages.

On the other hand, your raising of wages is often resisted by a union, and here you are in a good bargaining position. You may get *better* people and more of them from the BA if your option to raise wages would arouse resentment by workers who are *not* working for you; ultimately, this resentment would be directed against the BA.

The BA wants more wages for his trade, but friction results when some workers are paid more than others. If you represent an out-of-town contractor, you may interfere with work of local contractors by offering overtime work or a higher hourly rate on your job. You are draining the union members off the jobs where the BA has promised them. The BA will be unable to furnish people to the local contractors as he has promised, and the union will be at a disadvantage at the next contract negotiations.

Unlike other unions, the construction trades assume a responsibility not only to their members to keep them working, but also to the employers to have skilled workers available; many union agreements give the employer the right to hire his own people when the union cannot furnish them.

7.15 PARTIALITY TO CONTRACTORS

The BA wants to get along with everyone, particularly the union contractors he serves. The union is sometimes used by the contractors to control the market, preventing outside contractors from competing in the area. If the outside contractors are union-affiliated, this may be done by denying them *labor,* or by sending the least efficient people. By increasing construction costs in such a situation, both the union and the contractors profit. As construction costs rise, however, open-shop contractors (who may really be employing union labor from another city) find it more profitable to break into the market, and will do so if a job is large enough to warrant the anticipated trouble. A large contractor can ignore threats of boycotts which a small contractor cannot. The unions may also control prices by preventing entry of contractors into the business; this is rarely done in the case of general contractors, but does happen in the subcontract trades.

7.16 TRADE JURISDICTION

Although labor unions usually attempt to follow established customs in the trades, there are numerous decisions that result in higher costs. This may be because of substitution of one craft for a less skilled craft; that is, unions require that electricians rather than laborers unload electrical fixtures, although either trade could do it equally well. They require that plumbers install toilet paper holders on metal toilet partitions installed by sheet metal workers. Since the holders are shipped with the partitions and sold to the general contractor, this is an inconvenience; the cost of time required to find the holders may be much greater than the cost of installation, and as sheet metal workers do not ordinarily install partitions, it is difficult to find competent installers. As a practical matter, carpenters will often install both partitions and paper holders.

As pointed out later, jurisdictional decisions are often made without consideration of the efficiency of production. You will agree with decisions that result in award of the work to the trade that can do the work most rapidly (or in the case of the laborers, at the lowest wage cost) but will disagree with decisions that give work to people who are not skilled to handle it. The BA realizes this situation and will often cooperate as long as his union's claims are not affected; this means that in some cases you may do as you please as long as he doesn't know about it.

7.17 HIRING

A bricklayer superintendent, on arriving in a new city, called on the BA and asked to go around his daily rounds to the jobs with him to get acquainted. When the new superintendent saw a bricklayer who appeared to be more productive than the average, he asked the man for his name and telephone number. In this way he soon obtained a list of the best workers in the city, and when his own job started, had excellent production combined with good union relations.

Few people have both the nerve and personality to accomplish such direct action. You may be able to rely entirely on the BA to send you good workers, but you may also load your job with either the BA's political friends (for the BA, remember, is primarily a politician) or with people who have been able to get jobs only with the assistance of the union. You need at least one local person well acquainted with the local situation; sometimes a sales or manufacturer's representative is a good starting point. If you are hiring any people, try to get the widest possible publicity; by advertising for a carpenter foreman, and setting up a time for interviews that obviously will not conflict with working hours, you may learn a great deal about the industry in the city from these interviews. The foreman should be free of any tie with the union, at least in your contacts with him; a requirement that the foreman be a member of the local is not greatly restrictive if you look for your own foreman, through the newspapers or the local employment service. Employment services are less effective than direct advertising, because you usually want people who are already working. Often you may pick up a foreman who has been working as a small home builder or contractor, but who has retained his union membership; these people realize fully the value of a day's work. Anyone out of work doesn't want to wait for a job, and it is hard for a BA to insist that one of his union members be discharged after being hired. Most union rules can be enforced by a simple majority of the union; although hiring restrictions can be passed by a majority, they may be unenforceable unless the support by the members is overwhelming. Often the BA compromises by requiring merely that the person getting a job for himself report to the BA by telephone.

On the other hand, if the BA refuses to furnish specified persons, you can exercise your right of refusing people sent out (depending on the local situation) or by discharging people as fast as you get them. If you are the last superintendent to get people when nearly everyone is working, you may have to discharge quite a few persons in order to get a reasonably good work force; that is, to *go through* them. One of the quickest and best ways to hire is to contact other superintendents who are laying off.

Labor efficiency depends primarily on your ability to select efficient people, and this depends on a contact for advice. The foremen and their acquaintances are your best method: once production conditions are established on the job, a journeyman or foreman is unlikely to suggest someone whom he knows will not be satisfied under these conditions. The U.S. Taft-Hartley Act allows you to hire anyone, as long as he joins the union or pays union dues after going to work; this is a powerful lever, and some firms who do union work and are technically closed-shop *never* hire a union member. They hire nonunion, usually unskilled, help and require that the employees join the union later. This is normally practical only where not very much skill is required.

Many locals have rules restricting the hiring of members except through the union hiring hall, and these rules are particularly severe on outside contractors. Attempting to get the workers you want is therefore a constant compromise with the BA, who knows he has trouble enforcing these rules. The BA will often tell you who the better workers are, but he may want to break up crews to satisfy as many contractors as possible. A telephone contact with the other superintendent may get you entire crews with foremen that have been preselected on the other job. When there is going to be a general layoff shortly on another job, you may do well to postpone hiring–and delay your own job–in order to take advantage of the layoff. Of course, an acquaintance with the local contractors helps you to get the best people.

Your most important qualification as a superintendent is your contact with people who will work for you. You have always two conflicting obligations–to get the work done as cheaply as possible and to do it in such a way that your employees will willingly work for you. You should keep a record of addresses and telephone numbers of all your employees, with a notation of their ability. Very few contractors keep these records. You never know when you will be back in an area, although you may expect to do only one job there. Most firms have sketchy personnel records and put them in dead storage. It may be difficult to locate addresses and telephone numbers, and the office may have no record of abilities.

7.18 AVOIDING TROUBLE WITH THE BA

I have never seen a business representative go to a construction job to look for trouble–that is, to look for nonunion employees, nonunion contractors, less than prescribed rate of overtime, encroachment of jurisdiction, and the like. If a BA has no members on the job, he may come to find out why, or if passing he may drop in to stay acquainted with the people on the job. The job steward, normally your employee for a trade you employ, is responsible for the conduct of union affairs on the job.

The steward, however, has a rather limited interest in the union. The steward is not paid by his union but is designated for his duties, usually because he is a competent person. The steward usually may not be laid off, except for cause or during a layoff of all the people of his craft. The steward's duty is to keep the trade workers satisfied, and he is unlikely to take any action that would not satisfy the steward's co-workers. This is more important to the steward than union demands; the two are often not the same.

Sometimes there are several subcontractors on the same job, all employing workers of one trade. For example, both the roofing and air conditioning subcontractors may have sheet metal workers. The steward will be an employee of one of the subcontractors, but responsible for duties involving another. The steward's employer will not want to pay the steward for such union duties as checking cards of another employer's people. As a practical matter, the steward will not be active toward the other employer. If carpenters are installing metal toilet partitions, the roofers' sheet metal steward, who has never installed partitions, may not report them, for example. On the other hand, if the work is being done in the same room in which the steward is working, so he can't say he didn't see them, he may feel he must report the matter to avoid disciplinary action by his own union.

The business representative is normally a troubleshooter who comes to the job on the request of the job steward or by request of an outsider. A common cause of complaint is your discharge of a member who knows that working rules are not being followed (or at least feels that they are not), and who resents being laid off. If some other trade is doing work a worker believes belongs to his trade, he will think this is the reason for his layoff, and is doubly anxious to report the matter. Consequently, layoffs should be kept to a minimum, particularly if there is work that might be a source of contention.

Another source of complaint is the unemployed mechanic, who tours jobs under way either from curiosity or looking for work. A worker finds an encroachment on trade jurisdiction and complains to the BA, hoping to get a job. This person is not necessarily in a class with a real troublemaker; if he has a good reputation, it may be wise to put him to work. It may be that the work he noticed still proceeds as before.

Troublemakers, in this writer's opinion, are those who find infractions of working rules in trades other than their own. If the workers of a trade are satisfied, other trades should have no complaint. Sometimes a subcontractor's foreman is a self-appointed steward for all trades; a word to the subcontractor will usually get such a person to another job. These complaints may arise also because of personal disputes between foremen.

7.19 AVOIDING GENERAL STRIKES

You have two kinds of labor delays to contend with: *general strikes* and *walkouts.* General strikes against all contractors in a *bargaining unit* are settled by the unions and contractors' negotiators and are *called* over general issues, principally wages.

A dispute over matters on your job may not originate over your actions, as the walkout may be a subcontractor's affair. You can deal directly with the BA or steward on these matters. They occur over anything except wages—usually over jurisdiction.

There is a method to avoid a really important issue—wages. If your firm is not a member of the local contractors' association (that is, is not involved in the negotiations) or it has a national contract (made with national headquarters of the union), the union may agree to continue to provide workers for your job, contingent on your agreement to pay retroactively any increase in wages later agreed upon in the area. You obviously have nothing to lose by this arrangement, as you would have to pay this wage later anyway. The union is using your job as a place to keep its members working while negotiations continue, discriminating against its own local contractors.

Also, if you are not a member of the local subcontractors' association, you may continue work in specialty trades during a strike by putting the subcontractors' employees directly on your payroll. You are, in effect, taking advantage of the usual subcontract clause whereby you may take over a subcontractor's work and charge him the difference in cost to complete the work. In this instance, it is done by agreement, and since the same crew continues, there is no change in the subcontractor's cost. He is receiving part of his payment in payroll, and the subcontractor himself may continue to supervise the work. The union could hardly be fooled by this sort of strategy, especially since the subcontractor may continue to appear on the job. However, the union is able to keep the subcontractor's other jobs closed down, and they prefer to believe the legal technicality that they are not furnishing workers to the subcontractor being struck. Because of their independence, it is not possible to forecast definitely what a given local will do in any situation, even though the union's policy is definite.

Before taking any action different from that of other contractors in the area, consult your employer, because an important question of ethics arises in this situation. Contractors expect (or at least, hope) that other contractors will not undermine bargaining employers during a strike; by hiring under retroactive agreement, you are doing this. If a group of contractors are unfriendly to an outside contractor, they may not expect much support from the out-of-town firm during a strike.

7.20 RELATIONS WITH YOUR FOREMEN

As a foreman, you used a successful method of supervision. Do not assume *your* method to be the only one, attempting now to change *your foremen's* methods. Possibly new foremen may be trained in your own way, but it is unlikely that you will train present foremen to a new approach.

This does not imply that any foreman cannot improve his work, but points out that different people achieve success in supervision in different ways. Some are frank and direct; they can dish it out and they can take it. Some are restrained, and use an occasional word like a whip. It is important for you to develop patience—even patience to observe things being fouled up. It is worthwhile to criticize only the best people—the others may be laid off. And the best people are the ones you are most careful not to offend. So any direct criticism must be applied in carefully measured doses.

A new superintendent has a strong desire to supervise the mechanics, especially when he is unable to obtain foremen to do this properly. It is quite possible that you may be more valuable to your employer as a foreman than as a superintendent; but if you are a superintendent, there are other things to be taken care of than the foremen's work. If you attempt detailed supervision, you are quite likely to start your morning rounds and fail to return to the office all day. You are therefore entirely unavailable to other trades, who call the contractor; you will find you have again made yourself a foreman.

By watching a foreman and his workers, a superintendent gets an idea of his capabilities, but this is merely an estimate. A proper evaluation of a foreman's ability depends on a reliable estimate of the value of the work his crew does, to which the actual cost can be compared. The foreman must have the authority to hire and fire (or at least transfer) members of his crew, or his production has no meaning. In other industries supervisors are evaluated according to a rating score of personality characteristics, and the person with the highest score, according to the supervisor, presumably is the most efficient. Such methods often fail to take into account a foreman's outstanding ability, which may be important enough to outweigh all his weaknesses. When a foreman cannot be judged in dollars-and-cents terms, some other method is necessary; but do not judge him by his ability to conform to your own personality, either whether he uses the same methods as you do, or by how well he complies with your orders. The ultimate test is whether a foreman is making money; if he is, you must figure out some way to get along with him. *He doesn't have to respect you, like you, or perhaps even obey you; he just has to be right.*

You must select the best foremen you can find; but when the best have serious weaknesses, you have to accommodate yourself to these weaknesses. One foreman may get good production but be weak in following details, so you must constantly watch him. Another may be a good planner but will refuse to discharge a worker, regardless of how poor he may be. One foreman may find every possible excuse to avoid working with his tools, you may have to see to it that he works with a large crew to supervise, or not at all. Another insists on working, and with a large crew he does not properly supervise his people; he must work with a small crew. Depending on the size and type of your job, you must choose between the person who does all types of work reasonably well and one who does very few things very well. A foreman may be a good doorhanger but no good at all on formwork; on a small job, his abilities on finish carpentry may not be important enough to overcome his inability to handle other kinds of work.

It is often heard that "Workers can't be pushed any more." Probably they never could; the difficulty with the *carrot-and-stick* technique is that the stick is not very effective if another contractor is using a carrot. Your threats to fire a foreman (usually very quietly implied, but threats nevertheless) will not be effective if he knows he can get the same pay elsewhere with better treatment. But you must be very definite in setting forth minimum requirements; a foreman should know, for example, that he must turn in his time slip promptly or he will not remain a foreman. Likewise, safety rules are often viewed as of little importance by foremen. Although awards of stars and similar rewards for a low accident rate is helpful, it is occasionally necessary to make clear that failing to clean up nails or allowing open fires in buildings will produce more serious results than failure to win a badge or award in safety—such as being fired.

For initiative and efficiency, foremen must be led, not pushed. For enforcement of minimum regulations, equivalent to police regulations where little or no judgment is involved, it may be necessary to push them. This is particularly true in the area of safety; many otherwise good foremen lack sufficient imagination to see the consequences of unsafe practices, or are just incurable optimists.

Where possible, foremen should get a percentage of the amount they save under the estimated labor cost. To be effective, the labor estimate for this purpose should, more often than not, be one that can be reduced by the foreman. This bonus arrangement requires accurate estimating and accurate cost reporting and has a number of disadvantages:

1. The foremen are likely to waste material in order to save labor.
2. Workmanship may be slighted.

3. Friction between different trades may be created, as each is concerned only with his own crew rather than with the entire job.

These difficulties can be reduced by hiring foremen who have an experienced and mature outlook—those who would be capable of being superintendents on small jobs. It is also important that the foreman get a rate of pay that will be competitive with other employers, and that he not feel that he must make a bonus in order to make his full wages. He should expect a bonus on occasion but be reconciled to the fact that many jobs will not pay a bonus. He should not look on the bonus as something to be expected and on which he may borrow. A satisfied person must not be in debt, and therefore obliged to get a bonus, or have to look for a better-paying job. Although you may provide a base pay as high as does anyone else, there are always "pie-in-the-sky" jobs promising various special incentives. These look especially attractive to a desperate person.

Union rules and customs which require that supervision be exercised through the foremen are resented by some contractors, especially when local members must be hired as foremen. Sometimes the foreman has only a title, and the job superintendent directly supervises the crew. As pointed out before, there are a wide variety of people to choose from in any local, and there should be a good foreman available. The BA's choice is likely to be made for reasons other than ability; he may require support in the union of certain members and want to do them a favor. Some unions have no interest at all in the selection of the foreman or protection of the foreman against the superintendent; they consider the foreman a representative of management and require only that he be a member of union—but not necessarily of that local. This is particularly true of the bricklayers.

It is desirable that the foreman do his job, even though you may feel he is unnecessary. It may be that the hiring of extra foremen is most efficient, so that the size of each crew can be kept to three or four persons and the foreman can continue to work with his tools. For example, if you need 16 carpenters, and two foremen are required by union rules, you may be better off to hire four foremen and a *general foreman.* You can attract better mechanics for *leading men* and afford to pay for a better person for general foreman. In a study made a number of years ago, comparing the construction of identical ships at Newport News Shipbuilding Co. and the Portsmouth Naval Shipyard (which is almost directly across the river from them), it was found that the private firm employed many more foremen for the same number of workers, but the cost was less for the more closely supervised work. This type of work is comparable to building construction.

A good foreman is worth his wages, either as a mechanic or as a foreman; the difference in wages does not reflect the difference in production, and most foremen are underpaid in comparison with their efficiency. This was demonstrated at a Michigan shopping center where the electrical contractor used all his foremen on weekends at time and a half, bringing them from all his jobs to make crews for weekend work for the one job that was behind schedule. The cost of the work done was less for the premium overtime workers, at foreman rates, than for regular mechanics working during the week. Appointing someone as a working foreman may be nothing more than a way to recognize and to pay for superior production without antagonizing fellow workers.

7.21 DUTIES OF THE FOREMAN

A great deal of emphasis is often placed on the foreman as a "pusher," and his ability to get more work per dollar is his primary qualification. But he can encourage loyalty to the company—although this is usually loyalty to the foreman, who in turn is loyal to you. By *loyalty* is meant the employees' inclination not only to work but also to call the foreman's attention to the faults of other trades as well as their own. Criticism in craftsmen must be developed; they are accustomed to minding their own business and may disregard not only the inefficiency of others, but dangerous situations as well. You also want people who will remain on the job until it is finished and, if possible, will follow you to other jobs. Loyalty does not mean they will overlook your errors or fail to criticize you—rather, the opposite. Craftsmen sometimes delight in doing something your way when they are convinced it is wrong, just to prove it *is* wrong; your errors may be poured into eternal concrete with great enjoyment!

You must be direct and honest with your foreman, and he with his crew—but you are not being either direct or honest if a foreman is not being informed of what is going on, which is often the case. The foreman should know what is happening—progress, costs, hiring troubles, union troubles with other trades (he'll already know about his own), extras, and future jobs in sight. Rather than tell him only what he must know, he should not know it, and he should pass this information on to the workers. The earnings of salaried personnel, for example, are often considered confidential; a journeyman assumes that a beginning engineer is making much more money than is the journeyman, and this affects his own outlook and union wage demands. Union officials are occasionally surprised to find that their members earn more than do graduate engineers. A mechanic who is making more than a superintendent, which occasionally is the case, can hardly preserve an image of a management representative as a fat aristocrat.

Employees should be supervised through their foremen as a matter of good business, regardless of whether it is a union obligation. If your foreman is not competent, demote or fire him; don't try to bypass him.

Foremen need two qualifications—ability and ambition; rather than choose foremen from applicants, you may have to convince a mechanic to be a foreman. Some people know (usually from past experience) that as foremen they will lose the comradeship they enjoy and will gain a lot of worries they would otherwise avoid, and are therefore not anxious to take on this responsibility for the usual small pay differential. If a person is uncertain about his abilities, it is unwise to push a supervisory job on him, but if he has past experience as a supervisor you may have to persuade him to take a foreman's job again.

7.22 AUTHORIZING OVERTIME WORK

You may want to work at other than regular working hours in order to:

1. Take advantage of good weather.
2. Work equipment to the best advantage.
3. Push critical jobs that are holding up the work.
4. Get more help by premium payments.
5. Pay higher wages to selected people.
6. Work longer because of a shortage of labor.

The premium cost of overtime (that is, the added hourly wage rate) is determined by:

1. Union contract.
2. Job specifications.
3. Federal wages and hours law.

Union contracts are usually more demanding than the other requirements mentioned, requiring overtime pay (sometimes double pay) for over 8 hours per day, and with the greatest number of holidays.

If work is open-shop or you are not following union rules anyway, public contracts will still dictate pay rates. On U.S. government contracts, union contracts are normally used to write the specifications, but there may be differences. On state and county work, job specification rates (which are minimums) are sometimes lower than you have to pay on open-shop work in order to get qualified craftsmen.

The federal wages and hours law requires time and a half (150% of the base hourly rate) for over 40 hours per week, *not* 8 hours per day. Holidays are not considered.

Some union contracts require paid holidays, while stating that holidays worked must be at double time. You should see if holidays worked are paid for, *plus* double time. If you pay just double time, there is no reason to quit work for holidays. That is, if you pay for holidays *anyway,* then the added amount for working is the base pay, the same as any other time.

In some kinds of work, there is a long setup time and therefore a short day. It may be wise to work overtime to take advantage of good weather; the cost can usually be estimated and you should figure out if such a plan is really cheaper. On the other hand, you may have to work outside regular hours in order to make a 40-hour week; this is usually premium pay on union jobs.

The cost of equipment is largely *depreciation,* and in the case of earthmoving equipment and cranes, it may be cheaper to work as long as possible, since the cost of the equipment goes on whether you work or not. Sometimes you may rent equipment by the month, regardless of the number of hours worked; it is then to your advantage to work it as long as possible. In order to make long hours profitable for contractors, operating engineers often contract for time and a half for overtime in areas where other trades get double time.

The relative importance of some parts of the work has been pointed out before. The importance of a work item may have no relationship at all to the number of people working at it, and when a critical item requires few workers, the entire job can be completed earlier by a relatively low cost for overtime.

Most trade unions are successful in obtaining double time for overtime because the contractors have no intention of paying premium overtime at all, except as a direct increase in pay; that is, employers pay premium overtime when workers are scarce and must be drawn from another area or from other employers. The amount of work done for this added pay is not of importance; if time and a half were paid, they would have to pay more hours to provide the same premium. Double pay may therefore be readily conceded during wage bargaining. During times when workers are scarce, overtime is offered to get them. The only difference to the employer between overtime and an increase in the basic wage is that when work is again slack, he can readily drop the overtime and return to the basic wage; if he had raised the wage, there would be considerable complaint from employees when the rate was lowered.

When more people are needed, it is much more effective to give a small increase in the workweek with a guaranteed period during which the overtime will be maintained than to attempt to work a long week. For example, a 44-hour week with a promise to maintain the overtime for 6 weeks will get more workers, and get more work accomplished, than a 48-hour week with no such guarantee. Owners and architects have a tendency to consider overtime merely a method to get more work done per worker, and get more work accomplished, than a 48-hour week with no such guarantee. Owners and architects have a tendency to consider overtime merely a method to get more work done per worker, without realizing the difference overtime makes in both the size and type of the work force. If the owner can be persuaded to pay the premium overtime, the contractor will profit by getting more and better people, and by completing the work faster, which also usually results in lower cost. When the contractor and owner share the premium cost of overtime, the contractor may gain by his ability to compete with other contractors for labor in a scarce market.

In many cases, you will not be able to give some people overtime without giving it to everyone of that trade. Different trades do not compete with each other, though, and you may usually raise wages in one trade by continued overtime without arousing the jealousy of other trades.

Rarely will you give overtime in order to get more work from the workers you already have: you may have to pay overtime when others are doing it, to keep your help. If you are in an isolated location where it is impractical to get more people at any price, higher pay is necessary and is usually given as premium overtime pay.

7.23 DISCHARGING EMPLOYEES

Nobody likes to deprive another of his livelihood, and firing a person does this. But sooner or later, you must lay off practically everyone you hire; layoffs, like death and taxes, are inevitable in the construction business. You will also be obliged to fire people from time to time . . . the difference being that you fire someone when you still have work for him, but he is not satisfactory.

Regardless of the circumstances, you have nothing to gain by becoming angry with a person who must be discharged. If you are the type of foreman who bluntly criticizes his employees, don't criticize the person you are going to fire; it doesn't make any difference to the job, and can hurt you later. Superintendents are often angry at themselves for letting a person get away with sloppy work so long, or they aren't sure that the person should really be discharged at all, and they take it out by being blunt to the person being discharged.

There are two reasons that a person should be discharged without being told the reason. First, you may have a worker who is really a good craftsman but for some special reason is unable to do the work. He may be ill, mentally disturbed, have domestic troubles, or the available work may be something he dislikes but must do because that is all there is at the time. Or, there may be a conflict between an employee and the foreman or with another employee. On your next job, you may be glad to get the same person you fire now for inability.

Second, each person discharged is a bulletin you are issuing to others who are acquainted with him on every future job. If you charge someone with lack of ability, he will defend himself wherever he goes, and people will hear his side—you keep the people who speak well of your company and discharge those who do not. Under the best of conditions, you may get a reputation for unfairness. Others probably know what is happening and will pay little attention to one story, but if there are a number of the occurrences, the stories, however weak, will accumulate in their minds.

7.24 UNEMPLOYMENT COMPENSATION

An employee laid off for lack of work or for insufficient skill receives, from your state unemployment board, payment during time unemployed. This payment is based on the year's earnings before the last quarter. If a person is laid off in April, for example, the payment is based on earnings for the previous calendar year (January through December). The amount varies in each state.

These payments are made from charges to the employer. Your firm eventually pays the amounts received by employees it discharges, and retains a liability for as long as 2 years after an employee is discharged. If an employee quits or is discharged for cause (such as theft or refusal to work), the employer has no responsibility.

The explanation above is a simplification, as the employer always pays ahead, not for past charges. But he pays it, nevertheless, by adjustment in his rate based on percentage of the payroll at the time his payment to the insuror is made.

Some firms have definite policies on discharge for cause, designed to avoid part of this liability. If you discharge a person for misconduct and report him as laid off for lack of work (or for lack of skill), you may cause an increase in your employer's costs. From your own standpoint, it may be preferable that discharges be reported for lack of work, to avoid dissension and possible proceedings at which you will be called as a witness. In so doing, you are increasing the cost, not of your own job, but of others in later years, and this amount cannot be easily calculated (you don't know, for example, if the employee will ever draw unemployment compensation). You should conform to your company policy.

Workers who lack skill for the work, or who are inefficient, are still eligible for payments if discharged for these reasons. But if a worker leaves your job voluntarily, this should be noted, so that he may not later claim payments at the expense of your firm. It is a good practice to find work for personnel you are laying off, in order to avoid unemployment payments and to keep the good will of the other employees. If you recommend poor workers to another superintendent, you can of course expect the same treatment in return; but if you place good workers yourself, you can expect more employees to stay with you toward the end of the job when you cannot offer a long-term job and when you often need workers most. Bricklayers are hard to keep in the fall, for example, if they know they will be laid off and have to look for another job at the slowest time of the year.

An employee may not have the work experience (which must be over 3 months, sometimes 6 months, for private employers) to draw unemployment compensation at the time you discharge him but still will draw money against your employer later. Also, if you find a job for a person who is laid off, this doesn't mean that he may not draw compensation on your employer after being laid off again by someone else. The payments are complicated, and few employees or employers fully understand them.

Architects, Engineers, and Inspectors

8

Your job will usually be supervised by an architect or engineer who represents the owner and inspects the job to see that the work is being done in accordance with plans and specifications. The other persons with whom you deal, including the business agent, are acting as sellers; they have a strong incentive to do business with you, and to understand your point of view. The architect, engineer, or inspector represents the buyer, and it is therefore your job to please him.

8.1 THE ARCHITECT

If your job is supervised by an architect, he will be primarily concerned with the finish and appearance of the building; an engineer will be more concerned with the structural portions, such as concrete, steel, and excavation. This difference is not significant, however, since you may have either an architect, an engineer, or both supervising your job, and either may inspect the entire job. From a practical point of view they may be considered the same unless both of them are on the job. In such a case, the architect usually inspects appearance, and the engineer, the structure (mechanical and electrical engineers check the mechanical installations).

The architect will usually make periodic inspections of small jobs, and will have a full-time inspector on the job for large ones. Government jobs have a particular individual designated as *resident engineer* or *contracting officer,* the equivalent of the architect or engineer on private work, and it is important that you know the name of this individual. For many purposes, only the resident engineer may make a decision, and you must avoid following instructions given by several people, each of whom insists he has supervision of something or other. On public work, there may also be *design engineers* occasionally visiting the job; these people may actually be making the technical decisions, but these decisions are not binding on you unless the resident engineer approves them.

Often, you can find out the extent of a person's authority by asking the representative to sign an order for anything he requests. If he approves or disapproves something but will not give a written order or verification to that effect, it is probable that he lacks authority to make the statement or give the order.

As the architect is the buyer and you are the seller, your problems come in two categories: you want to sell him something he won't buy, or he wants to buy something you don't have. That is, you want to get acceptance of some work, and he wants work done that he may or may not agree is beyond the contract requirements.

8.2 STUDYING SPECIFICATIONS

As soon as possible after being assigned to a construction job, you should completely study the plans and specifications. Don't hesitate to ask anyone available the meaning of words you do not understand. Pay particular attention to the portions of the specifications relating to your duties and to the duties of inspectors, architects, and engineers. The *A/E* (architect/engineer) is limited in authorizing changes by the owner, but it may be that this authority is given the A/E by the owner, not to you. You should *promptly* report to the contractor any extra work, interpretations of plans, or changes by the A/E.

8.3 OBTAINING WRITTEN AUTHORITY

Luckily, it is not your responsibility to present your employer's case to the architect, or if later necessary, to an arbitrator. Your job is equally important, if not more so—*to get the information on what actually happened, to get a written order from the architect, to keep clear records on weather conditions, strikes, and the like. You need specific and written information on when material arrived—information to which you can swear in court if necessary.* Most extra claims and claims for time extensions get confused because neither side is really sure what actually happened, and when.

As far as you are concerned, the architect's authority (but not that of his inspector) is final. He can tell you to put work in, to take it out, to start it, to stop it. He can require any amount of extra work he cares to order. He may or may not be required to pay for the work he orders, but he can be required to sign a statement directing what is to be done. Such a statement should be simple and direct, as, "Remove east wall as directed," or "Insulate hot water piping in Room 204." Do not accept vague statements such as "Hang door in Room 458 in accordance with specifications." Some examples of good and bad written orders are:

Good

"Provide drain from basement to east ditch as verbally directed."

"Repaint Room 354 with Sherwin-Williams Green No. 35 in lieu of existing No. 32."

"Install toilet paper holders in all toilets."

Bad

"At no extra cost, extend footings to 4 feet below sidewalk."

"Insulate all piping in accordance with specifications."

"As previously directed, install door as shown on Addendum No. 1."

The "good" orders above are definite, and in case of the painting, establishes the work to be covered or to be removed—but be sure that the paint previously used *was* No. 32! Verbal instructions are acceptable to provide details—there is usually little trouble later about what was to be done. The first "bad" order above lists a condition—*at no extra cost.* If you know there is really no extra cost, this order is all right; but if there is an extra cost, you may never collect for it if you comply with such an order. The architect can order work to be done, but the determination of whether there is an extra cost is a matter for the contractor to dispute. The second order above, to insulate piping in conformance with specifications, just doesn't say anything unless there was some discussion as to whether the insulation was to be omitted or if the pipe was ready for insulation. It does not say *what* pipe is to be insulated, but has been used to supposedly enforce a ruling of the inspectors that the specification did require insulation of certain piping. The phrase *as previously directed* should never be accepted unless you were, in fact, previously directed. By proceeding with work under such a qualifying statement, you may be admitting you did not do something when you were supposed to do it.

When you have established facts by your daily report or otherwise, most of the disputes over extras can be readily settled. When extra work is needed, there are two orders necessary: one to do the work, the other to pay for it. If the work is needed, the architect should tell you to do it—*in writing*—and negotiate the matter of payment with the contractor later.

As you become acquainted with the architect and with your employer, verbal orders become routine—you know when to comply with them. Some architects refuse to give a written order for work they consider to be included in the contract; if you proceed without their approval, as in covering up work, the cost of later correction may be several times the cost of original omitted work. The architect still has an option to accept the work you install, should a later arbitration award be in your favor. For example, the architect tells you to insulate certain piping—without a written order but with the implication that it is included in the contract. You refuse to do so without a written order, covering up the piping with cabinets. If on later appeal an arbitrator decides that the work is included and should be done, you must do much more than the omitted work to rectify the situation.

On the other hand, if you *do* the work and claim an extra, the architect may deny that he ordered any extra work at all: that he ordered work only in accordance with the specifications and can't help it if you thought this work was included in the specifications. In this case, you may not collect for the extra work, even though your interpretation of the plans was correct.

This is not to imply that architects are a shifty lot—on the contrary. Most architects are quite careful to give the contractor all that is due him, and will on occasion themselves pay for work necessary because of errors in their own plans and instructions—which legally they are not obligated to do. You are most concerned with the occasional difficult architect, and in many cases the safest thing to do is to stop work on the disputed matter and call for help from the contractor. It's his money, and he's entitled to assess the risks. Don't say you're stopping work on the item unless this is obviously what the architect will approve, or don't threaten to stop work—you have no legal right to do so. Also, do not attempt to demand a concession that certain work is included in the contract before you start it—this is beyond your legal rights. By stopping or refusing work you may be jeopardizing the contractor's rights under the contract. If you proceed under verbal orders, attempt to get the orders in someone else's hearing and get an immediate signed statement from others as to what was said, containing the *exact words* used. A later statement that, "He said to go ahead with these joists, but I don't know for sure just what kind he said to use," is not sufficient.

Also, immediately (in construction this word does *not* mean "as soon as possible") write a letter to the contractor giving your version and understanding of the oral order, including facts such as names, dates, and even time of day and, if true, that you are proceeding "under protest." Do this even if the contractor himself was there when the order was given. Sometimes you will be the only one who has the "feel" that what appears to be a minor decision may later turn out to be important. In this case write a dated memorandum of the facts to yourself and file it. Letters and memorandums like this may later become "gold in the bank," when the contractor is obtaining final settlement of the contract.

Remember also that if an extra was ordered, the contractor may not collect for it if he fails to claim the work as an extra *before the work was done.*

8.4 INSPECTORS

The A/E may be employed by the owner to prepare plans, to supervise construction to assure compliance with the plans, or both. If the A/E does not supervise construction, you will probably not see him or his representative at all.

The A/E inspector checks compliance with the plans. Since municipal codes and other code requirements are included *by reference* in the job specifications, he will probably check work at least partially, in accordance with these codes. Primarily, on private work (nongovernmental), other inspectors check the requirements of their own agencies. As far as your job is concerned, you have authority to represent the contractor. What you actually may do is directed by the contractor, but if you commit the firm further than you should, the firm is usually bound to carry out your promise. Your authority to accept orders is not restricted; you are obliged to relay any verbal or written instructions the definite status; often no one, including themselves, knows just what they are supposed to do and how much authority they have. They are something less than agents of their employers and something more than sidewalk superintendents. The inspectors you will usually have on your job represent the architect, and may be *architect's inspectors* or, if the job is publicly owned, may be *federal* or *state inspectors.* They are there for the same purpose—to protect the owner's interests, and to assist the superintendent when interpretation of plans or specs becomes hazy.

There are *municipal inspectors* and/or *county inspectors* to enforce local building codes. These inspectors will usually be general construction, electrical, and plumbing specialists, and are often retired mechanics in their own trades. A municipality may also have *fire inspectors,* who check both on compliance with safety regulations regarding storing of inflammable materials during construction and on features of the permanent construction about which they are concerned. This includes provision of proper threads on any fire department connections and provision for exit of people and entry of firefighters. They may insist that hardware be installed on back doors, for example, which will allow them to get into a building in case of fire; they are interested in proper venting of smoke, and installation of panic-door hardware in auditoriums. There may be special inspectors separate from the plumbing department to check on public health facilities such as septic tanks and swimming pools. Specialized buildings like hotels or schools may have inspectors from special agencies set up to enforce laws pertaining to that kind of building.

Safety inspectors may inspect your job to see that regulations concerning provision of proper scaffolding, grounding of tools, and other safety features are observed, as further explained in Chapter 12. These people may have the power to assess fines for noncompliance with regulations. Inspectors from your insurance company may also check the job, particularly if it is a large job with a high accident rate.

Underwriter's inspectors are less noticed because they may not see you at all, and infractions of their rules may incur an increase in the permanent fire insurance rates rather than any immediate notice or cost. They inspect installation of sprinklers, fire doors, fire walls, and other work that affects the ability of the building to withstand fire.

The *mortgagee* or *mortgage insuror* may have his own inspector also. The mortgagee (*lender:* usually a bank, savings association, or insurance company) depends on the other inspectors for technical compliance. His representative may check the job before it starts, to be sure his mortgage is recorded before any work is done, and he will have a surveyor check the location at the end, and sometimes during the progress, of the job. He will check the progress of the work against the owner's or contractor's payment and in comparison with the total contract amount. The mortgage insuror (FHA or VA) will make regular inspections of insured work.

With the exception of the safety inspectors, all these people are checking the job to see that construction is in accordance with plans. If the plans are definite, you have little trouble with them; the more vague the plans are, the more details must be agreed upon by yourself and the inspector concerned.

All these inspectors will rarely be on the job. On a government job, for example, the government inspector replaces all other inspectors, since government work is not subject to municipal or state regulations, nor are there insurance or mortgage requirements.

The *architect's inspector,* on private work, is the "eyes" of the architect. His power is largely negative, as he may refuse work, and his acceptance is not final. Even his power to refuse work is limited and, if the conduct of the inspector (particularly when he refuses work with resultant delay in the job) is objectionable, the architect should explain what the inspector's authority is to be. If the architect chooses, he can delegate full authority to the inspector.

Federal inspectors on government-owned work and, to some degree, state inspectors on state work have powers entirely different from those of private inspectors. They are not agents of the government in the sense that a private inspector is an agent of the architect. Even the authority of the contracting officer is limited by statute, and the best and most friendly contracting officer can do nothing to get extra payment where there are statutory provisions to the contrary. Usual principles of law often do not apply against the government, and the authority of public officers to order work may be severely limited. For example, a government inspector required a contractor to place rock in a way the contractor believed to be an extra to the contract. The contractor's view was later upheld to be correct, but he had not appealed to the resident engineer as soon as he started the work. It was therefore held that the contractor could not collect for the extra work done *before* he claimed extra payment, but would be paid for identical work completed *after* the claim was made!

8.5 INSPECTION CHECKLIST

Most inspectors will come to your job without being called, or will be called by the architect or contractor. Sometimes the specifications require that certain certificates (such as a roofing bond) be furnished to the architect before final payment is made. Although these inspections are usually ordered from the office, you should call the contractor when you are ready for inspection. You must call for inspection on any of your activities that might harm others—such as burning trash or barricading part of a street.

Many inspections required by local regulations are not actually made. For example, a large shopping center was recently built in a major metropolitan area. As the architect was hired only to prepare plans, he did not have an inspector on the job. The city electrical inspector usually merely checked in at the electrical superintendent's office and often did not actually look at the conduit being covered before a pour. The general construction inspector not only failed to see the reinforcing steel before concrete was poured, but halfway through the job came to the superintendent for a plot plan—the inspector did not have drawings of the structure, and he had no plans showing the number of buildings on the site; he explained he wanted a drawing so that, if asked, he could at least find a building by its name! This example is not an indication of negligence; local officials are concerned with smaller structures which are not A/E-supervised and often not designed by A/E's. Municipal officials rely on A/E's to properly design and inspect construction on which the A/E's are engaged.

Necessary inspections required you to:

1. Check the specifications thoroughly for any indication of inspections, particularly provisions that require certificates that may not be readily obtained after the job is completed.

2. Contact the agency granting the building permit for a list of required inspections.
3. Ask each mechanical contractor, before you cover up his work, if all required inspections are complete. This is particularly important if the subcontractor is working on the job for a few days at a time. The foreman may complete his work, which is otherwise ready to be covered, and forget to tell you an inspector is expected later.
4. Learn which inspectors insist on seeing the work and which allow you to cover the work without inspection if they are late. Many inspectors will ask to be called a certain number of hours in advance of a concrete pour, but will allow you to go ahead with the work if they are not on the job in time.
5. Call the local fire department to find out what inspections and permits they require. Since you do not usually have a building permit from them, they may not know you are starting work.
6. Ask the architect or engineer for the list of inspections he is to make, or ask him for a list of any inspections required by others. If the contractor is to hire an independent testing agency, as is often the case, arrangements must be made for its inspections. Inquire particularly about inspections by mortgage holders, FHA or VA, tenants, or state regulatory commissions.
7. Call the county or city health department for regulations regarding your temporary toilets, as well as permanent water supply, sewage disposal, and swimming pools, and ask when inspections, if any, are to be made.
8. If you must get any permit or authority—even by telephone—to do any work on public property, such as streets, to obstruct traffic in other ways, or to connect to city utilities, ask about inspections—if, by whom, and when.
9. If you are using explosives, notify the local law enforcement agency, and discuss shooting schedule, hauling and storage arrangements, and any safety precautions affecting traffic on public streets or roads. Mobile radios, for example, can be dangerous near blasting, because blasting wires act as antennas and pick up electrical currents.

8.6 COMPLIANCE OF CONSTRUCTION PLANS WITH CODES

Under the standard agreement of the American Institute of Architects, a contractor is not responsible for checking drawings for compliance with building codes, but he is responsible for following the code if he knows there is a conflict. He is entitled to extra payment for this work. Some specifications, however, require the contractor to conform to local building codes without extra payment.

In either case, you should check plans for code compliance to the extent of your ability, and ask subcontractors to do the same. You should never proceed with work you know does not comply with local building codes, except with authority of your employer. You may proceed with corrected work without specific orders by the architect and still collect for extra payment, but you should do this only if there is no doubt as to how the plans must be changed. In most cases, the change in plans is not at all obvious and you should wait for the architect's instructions. For example, increasing a partition from 4 to 6 inches in thickness may be a very minor matter; or there may be so little clearance in the building that it would be necessary to change the type of wall construction or increase the size of the entire building. On the other hand, it may be required only that a few courses of concrete block be filled with concrete, an operation that was not indicated on the plans, in which case it may be readily done on the job without delay. Public work may not have to comply with building codes, but you should check with local building authorities regarding any discrepancies.

8.7 DUTIES OF THE LAND SURVEYOR

The owner furnishes the corner stakes of the property and provides a drawing showing these corners. This drawing should be complete, and the stakes should be so marked that there is no doubt whatever about their location. It is better to go over the property with the owner, architect, or surveyor, put a personal marker of your own at each corner, and ask him, "Is this the corner shown on the map?" This is the time to be really stupid. Don't accept any instructions that a corner is "in the bush" or "5 feet from the stake." There is a great deal more to surveying than merely measuring distances and angles; the owner's lot may not be the size shown on the plot plan, subdivision plot, or deed. Make him show you the exact point; it is not uncommon to find several pipes or stakes near the true corner, and there may be a disagreement among surveyors. Never, *never* set a property corner marker yourself; many contractors hire a land surveyor to set the building corners as well as property corners. There are very few mistakes that can make a contractor go broke on a single job, and of these, the only one you can completely control is the location of the building—you must put it on the right place! On important work you should have a surveyor check the location of the foundation walls as soon as they are built.

If you must remove the boundary markers, set reference points in the presence of responsible witnesses (preferably the job inspector) or have it done by a land surveyor.

8.8 CONCRETE TEST CYLINDERS

Sampling of the work for test purposes is usually done by the job inspector. There is one important exception to this, however—the sampling of concrete. It is often the contractor's responsibility to pour and cure test cylinders. Specifications typically require that a set of three concrete samples be made from each major pour or each day's work. One of these is tested when 7 days old, to give a report as early as possible; at this time, concrete develops about half of its 28-day strength. The second cylinder is tested when 28 days old, and the third cylinder is kept in reserve. If one of the first two cylinders should be unsatisfactory, the third cylinder may be broken as a check.

The test results from concrete cylinders vary greatly. Test results are changed by the manner of placing the concrete in the test form, the place or load from which the sample is taken, and moisture and temperature conditions. Consequently, if an inexperienced inspector takes a test improperly, the laboratory will report the concrete as unsatisfactory. When an architect is making occasional visits to the job rather than maintaining an inspector on the job, he will usually not be present when cylinders are taken. Both to protect your employer and as a convenience to the architect, you should be able to do this work yourself.

If the test cylinders prove unsatisfactory, cylinders may be cut out of the completed building; if these also prove unsatisfactory, the contractor may be faced with the necessity of making a full-load test on the building or replacing portions of it. The concrete supplier may be responsible for the cost of such work, but only if he has previously agreed to make such a guarantee. Many concrete suppliers will pay for such work, although it is often hard to tell if the supplier is to blame.

If you are to make these tests, call your concrete supplier for instructions. He will send a person to show you just how it is done; if he has no one available, he will have a test laboratory send a technician. Don't rely on telephone instructions, written instructions, or an explanation using the equipment on a "dry run" (without concrete)—make several cylinders under his supervision, using actual concrete. The operation is simple, but a failure can have serious consequences. The concrete supplier is greatly interested in having proper tests made, as he can lose payment or be responsible for further expenses if the tests are not made properly.

Curing is as important as testing the strength. It is rarely possible to cure a small sample of concrete under the same conditions as the concrete in the building, so you should cure your samples under the best

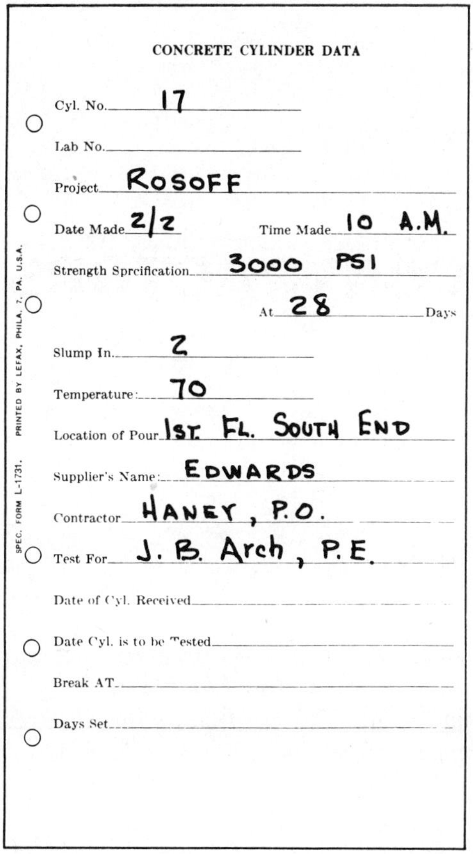

CONCRETE CYLINDER DATA

Cyl. No. 17

Lab No.

Project ROSOFF

Date Made 2/2 Time Made 10 A.M.

Strength Sprcification 3000 PSI

At 28 Days

Slump In. 2

Temperature: 70

Location of Pour 1ST FL. SOUTH END

Supplier's Name: EDWARDS

Contractor HANEY, P.O.

Test For J. B. Arch, P.E.

Date of Cyl. Received

Date Cyl. is to be Tested

Break AT

Days Set

SPEC. FORM L-1731. PRINTED BY LEFAX, PHILA. 7, PA. U.S.A.

Figure 24 Concrete cylinder data (Lefax).

possible conditions—usually, a box of wet sawdust maintained at the proper temperature. After curing, samples are sent to a laboratory for testing, with an identifying label as illustrated in Fig. 24. If number and other identification is marked on the cylinder itself, a written label may not be required.

Cardboard forms may be used for making these cylinders, or if there are many to be made, a reusable metal form is used.

Subcontractors

Bricklayer foreman to electrical foreman: "You'd better get your boxes in that east wall—we're starting blockwork."

Electrical foreman: "But I've got two people over there now."

Bricklayer foreman: "Yes, but they're drinking coffee and my people are laying block."

9.1 SUBCONTRACTOR ORGANIZATION

In building construction the greater part and sometimes all the work you supervise is done by subcontractors. The sloppy superintendent welcomes subcontractors because he leaves them alone and is happy to use their slow progress as an excuse for failing to finish his own work. The careful superintendent dreads subcontract work because without direct control of the labor he cannot control overall progress, and without detailed knowledge of the subcontractor's material on hand and ordered, he doesn't know if the subcontractor's delay is unavoidable.

Large organizations which themselves must lose money because of delays in completion of their work let as little work to subcontractors as possible, to better retain control of progress. Architects and engineers often specify a maximum percentage of work that may be sublet, for the same reason. Why, then, is work sublet at all?

The subcontractor must be able to furnish a working force, supervision, materials, and sometimes credit which is not available to the general contractor. The general contractor rarely does enough of one type of work to continuously keep workers and supervisors active in many trades, and his quantity of material purchases of special items is too small to get the best prices.

Since the subcontractor is selected by the contractor, there is little that you, the superintendent, can do but trust that the subcontractor can do what he has been hired to do. If he cannot, you are interested in getting rid of him at the first opportunity; or if you must, you are interested in putting enough pressure on him—taking over his work force if need be—to get the work done.

9.2 CANCELING A SUBCONTRACT

You will not usually have to decide if a subcontract should be canceled because of the inability of a subcontractor to keep up or to complete the work properly, but your recommendation will be important, and sometimes you will be the contractor's only source of information. If a subcontractor has been slow, he usually knows it. If the contract was properly drawn, his failure to keep up should be clear and there should be no argument. However, will cancellation help you complete the job more quickly or cheaply? You can tell from the subcontract if you can readily cancel it and, if so, what steps you may take to complete the work.

If you have a clear description in the subcontract of the work the subcontractor must have completed on specified dates, or the number of workers he must have on the job at certain times, you may cancel the contract when he obviously fails to comply with it. If the requirements of the subcontract are vague, you need to establish such dates as far as possible. For example, there may be no mention of progress in the subcontract, other than a vague reference that the subcontractor is liable to the same extent as is the general contractor. There may be a requirement that the subcontractor is to "keep up with all trades," "not to delay the work," or "to complete various phases of the work as directed by the superintendent"; these are all strong clauses—so strong as to probably be often unenforceable. If the contract contains an impossible requirement, you may not be able to claim anything at all under that clause, and it may even invalidate the contract.

It is rare that an informed subcontractor is unable to meet a schedule made by an experienced superintendent; more often, the subcontractor merely refuses to provide a reasonable schedule, and refuses to state definitely that he can or cannot meet the superintendent's schedule. *Consequently, as soon as you get on the job, you must start keeping records of items that may be cause for cancellation of the subcontract; otherwise, you will be unable to justify cancellation if it becomes necessary.*

First, a date is set for the various parts of the subcontractor's work as soon as ground is broken; this may be 2 or 3 months in advance of the actual work. About a month before he is due on the job, write him, referring to whatever contract language is available and confirming the date set earlier or setting a new one. A week or so before he is due on the job, call him and send another letter confirming the call. If he has a small contract for which labor is assigned on a day-to-day basis, call him again 1 or 2 days before he is due. It is immaterial whether he agrees to comply with these dates or not; if he has a good reason to delay the work, the dates should be changed, but his general statement that he hopes to get to work on time may be ignored. By the time he is actually due on the job, he will expect he really has a real stickler for a superintendent; you can then be friendly and reasonable—the old "hero and villain" routine. When the subcontractor arrives, insist that his foreman have full authority to decide with you what work is to be done first, and that no workers are to be removed from the job without your consent. This will protect the subcontractor's foreman from being overruled by his office when he promises something will be done at a certain time.

Removing personnel from the job should be interpreted as the subcontractor's stopping work under the contract—most subcontract provisions can be stretched to this interpretation. Make it clear from the first that you are friendly—you'll buy him beer and pull his truck out of the mud—*IF* he does what he is supposed to do. If he starts to give another general contractor workers in preference to you, do a Dr. Jekyll and Mr. Hyde bit—all of a sudden you're a highly nervous, angry, and irresponsible person, not entirely rational, who wants nothing more than to get rid of the subcontractor for good. If your legal position isn't very good, you may have to ham it up a little more.

The subcontractor, when faced with demands of competing general contractors for his workers (which is usually the reason for his lack of progress), has two thoughts—first, which superintendent *can* legally cancel his contract, and second, which one *will* cancel, regardless of consequences. If your legal position is good—that is, if your case is well documented and your contract is tight—you need scare the subcontractor only enough to send him to see his lawyer. His lawyer will tell him, usually, not only to conform to the legal requirements but to avoid the "gray areas" of what neither of you can prove. As a practical matter, it is very difficult to prove lack of progress even on a well-written subcontract, and the subcontractor's lawyer will be the first to tell him this.

However, if you are going to threaten cancellation, you shouldn't bluff—you must be prepared (or rather, your employer must be prepared) to cancel a contract even if this means losing money. You have a reputation to maintain—for truthfulness. Cancellation of a small contract may lead to prompt completion of much larger contracts by other subcontractors.

When a subcontractor does a good job—and most of them do—send him a letter of appreciation and make it good. You may have fought with a subcontractor throughout a job, both for money and for time, but, when it's over, send him a letter that he will be proud to hang on his wall. Other people in the business will see it, and will get the impression (1) that you appreciate a job properly done, and (2) that some things they hear about how nasty you are (which is inevitable if you do your job) apparently aren't so. If your demands on subcontractors are clearly understood, some of them will like you and some won't—but few will think you're wishy-washy, the attitude subcontractors dislike most. If you step on one person occasionally, he knows that when someone gets in *his* way, he can count on you to step on the other person, too. If you're pushing the electrician in front of the lather, the lather doesn't have to push him. The lather stays on good terms with the electrician, but gets his work done—you're the villain and he can be the hero.

9.3 THE SUBCONTRACTOR'S SCHEDULE

Regardless of whether completion dates have been set up in the subcontract, you will need to get the subcontractor's agreement as to the time that certain parts of the work will be done. Few foremen (and you should be dealing with the job foreman) will habitually overestimate the time they take to do their work—usually they will either be overly optimistic or will refuse to even guess at a completion time.

There are several reasons why foremen are inclined to estimate that work takes less time than it does. They are superior craftsmen themselves or they wouldn't be foremen. Since the only way they have to estimate the time something takes is to figure how long they would take to do it themselves, they will not allow enough time for the less-skilled persons who are actually doing the work. Foremen are inclined to think out the operations involved, failing to take into account the waste motion and faulty work—which occur because the foreman isn't watching. Since he prevents these time losses while he *is* watching, his mental picture of the job often doesn't include them. You will discover, if you haven't already, that all of us are prone to remember the good production days when estimating time required. The bad is pushed aside in our memories, also resulting in optimistic estimates.

If the subcontractor's foreman is consistently optimistic—that is, he uniformly estimates he will finish items the same number of days earlier than actually happens—you add several days to his estimate. You are in a good position to insist on more workers later if you have accepted his date and have not pushed him for several days afterward (since you didn't expect him to be through anyway); it is obvious to both of you that the subcontractor is not meeting his own estimates.

If a foreman entirely refuses to estimate the time of completion, you must do so. You can set a date that appears reasonable, give it to him in writing—there is usually some justification in the subcontract for this—with ample time for him to answer. You can follow this procedure if the subcontractor's foreman gives you information on what is reasonable to expect but refuses to promise it. Give this estimate to the subcontractor as an order: the subcontractor will find he can do it—and wonders why you are such a good estimator!

9.4 STARTING TIMES FOR THE FOLLOWING SUBCONTRACTOR

You must get the completion dates from a subcontractor one way or another—by agreement or by his lack of argument. If the subcontractor says he can't complete some work at a particular time, ask him to write that down and sign it. If he won't write what he'll say, you have every reason to believe he is deliberately refusing to tell the truth. On the other hand, *you should be willing to write what you say.* If your employer objects to your doing this, you should just not talk so much. Don't ask subcontractors to work from your verbal orders if they have every reason to think you will deny them later. If a subcontractor admits he cannot complete some work, then under most contracts you have proof of his inability to keep up. So he doesn't have to promise he *will* get done at a certain date, but on the other hand, he can't say he can't get done or that your request is unreasonable, without putting himself on the spot.

You must judge a subcontractor's promise for what it may be worth, and then make a commitment of your own to the next subcontractor as to when he may start work. In this process you will make some mistakes, and it is quite possible the mistakes will cost your company money. For example, you tell a lather to start on a wall on a date which is a little later than the electrician has told you he will be through. The electrician does not finish, so you tell the lather to cover him up. If your subcontract is not well written or you lack adequate proof the electrician is behind, you may be unable to charge the electrician with later repairs; there is always a possibility that your action will cost your employer money. But should you fail to send the lather ahead, he will have no confidence in you thereafter; he won't come to the job when you ask him, but will wait until he can inspect the job and see that you're ready—and then may take several days to get on the job. The subcontract—and your own dates —should be such that the electrician has to worry about getting covered up, and to pay for the damages if he does; but you must be stubborn in any case.

You will find that if you are willing to cover up a trade, it will rarely be necessary: if subcontractors know you make your concrete pours on schedule, regardless of who is ready, the subcontractors will shuttle workers from some superintendent who is willing to wait longer for them. Also, the subcontractor is on your side in one respect—the foreman on the job, who gets covered up, doesn't have to explain his own lack of planning ability (which he wasn't hired for anyway), but he has to explain why his workers are slow and therefore inefficient. The foreman who tells his subcontractor you are unfair is fighting an uphill battle. This may make you unpopular at times with a subcontractor's foreman, but it gets the work done.

But this can work to the advantage of the job foreman as well. For example, a plumber calls his office for more workers to meet your projected pour. His office is getting several such calls—some of the foremen will be told to wait. But the foreman on *your* job will say, "This guy won't wait, and I won't ask him. Yesterday an electrician just got out of the way in time, but still got concrete on his shoes; they didn't wait for *him.*" The subcontractor won't believe what you tell him—but he'll believe his own foreman. Then if the foreman does *not* get his workers, and *does* get covered up, the foreman has a perfect excuse for the later extra work required—he tells the subcontractor, "I told you so."

You must take the same attitude with your own foreman if you are going to have a reputation for fairness. You must be prepared to pay overtime to get your work done on schedule, or to do special work (such as cutting reinforcing steel on the job) if you are going to convince the subcontractor that your schedule is more than a hope. If you set a date for a subcontractor to start and are unable to meet it because of your *own* forces, you can expect even less future cooperation from him than if another subcontractor held him up.

To insist on compliance with schedules does not mean that you attempt to complete the job in a hurry. The schedule may be an easy one, but you still need to watch the work carefully if you are going to meet it. To keep to a schedule, you should allow generous time originally. But once having established a schedule, it must be followed, or you can easily have a job with no progress at all. The cyclic nature of the trades should be established; the work forces should be adjusted so that all work moves evenly. This is necessary both for economy of supervision and for efficiency of labor. *It is easier to comply with a schedule than to nearly meet it.* To "nearly" meet a schedule allows everyone involved to form their own opinions of when they should be done. The schedule then becomes a constantly changing collection of the opinions of subcontractors and foremen, rather than a series of operations.

9.5 COMPLETING A SUBCONTRACTOR'S WORK

The terms of the usual subcontract allow the general contractor to cancel the subcontract if the subcontractor is unable to complete work on schedule. Cancellation often requires a great deal of negotiation, particularly if the subcontract is one such as electrical or mechanical work, for which a specialty license is needed. If the subcontractor is financially sound or if there is no doubt that the balance remaining on the contract is enough to complete his work, you may immediately proceed with your own labor or with another subcontractor on a fixed-fee contract, and no negotiations or bids are necessary. Under some contract conditions, you may retain the subcontractor's workers and even his supervisors. Most subcontracts allow the general contractor to keep the materials delivered to the job, and many allow him to take over the subcontractor's equipment. If a cancellation is expected, however, the subcontractor may move his equipment off the job before you can prevent him.

To continue work promptly after cancellation, you must confirm all orders placed for the job and obtain delivery of materials in transit. Otherwise, lack of materials may delay the job for some time. Because of the necessity of keeping the subcontractor's material, equipment, and workers, as well as the desirability of avoiding litigation, the cancellation should be by agreement with the subcontractor if possible. The subcontractor realizes he has little chance of collecting further money, and the general contractor knows he has little chance of collecting damages from the subcontractor, even though this may be allowed in the contract. If a subcontractor is late in his work, an offer to negotiate a settlement of his contract midway in the job may have surprising results in his progress; he may believe you're bluffing about cancellation of his contract, but an offer to negotiate makes it appear that considerable thought has been given to the matter.

The subcontract may allow you to cancel part of the work and get another subcontractor on the job; that is, it may give the general contractor an option of canceling portions of the work without relieving the subcontractor of the responsibility for continuing work. This makes a negotiated settlement unnecessary; you merely call on another subcontractor, on either a lump-sum bid or on cost-plus basis, to do a portion of the work. You may even engage a second subcontractor merely to furnish personnel who will work under the first subcontractor's direction. This will often be to the first subcontractor's advantage, as compared to his losing the work entirely. The first subcontractor still continues the work he must pay for.

9.6 TWO SUBCONTRACTORS ON THE SAME JOB

There is a reluctance on the part of some subcontractors to work on the same job as another subcontractor of the same trade, or to take over another's work. On small jobs, it may not be possible to find an agreeable subcontractor. On larger jobs, there is usually little difficulty, particularly if you make it clear beforehand that other firms will do such work. You may have a situation in which no established contractor will take over the work. In the case of electrical and plumbing trades, it may be necessary to send an inquiry to every person who has a master plumber's or electrician's license in the city; usually you will find someone to cooperate. A person with a *master's license* but with no money to finance a job may agree to supervise the work for as little as 1 or 2 percent. Such supervisors may lose their license by "renting their license" to another, but it is legitimate for them to work for salary if they actually supervise the work. The inquiry itself will immediately reach everyone in the trade (even though you do not send it to everyone) and may have a beneficial effect on the subcontractor still working on the job. In general, the subcontractor will put more reliance on information he gets from other sources than on what you tell him. There is a great deal of bluff and "eyeball to eyeball" discussion—the situation may be that either you or the subcontractor will lose money if the contract is canceled, but you may have to convince him that you're too dumb to know this. He will be doing the same thing, often resulting in highly emotional discussions and threats which neither party intends to carry out. Ideally, you should force him to do what you are considering—to call on other subcontractors for help. He can negotiate with them much more efficiently than you can.

9.7 SYMPATHY FOR THE SUBCONTRACTOR

Nowhere has it been suggested that the subcontractor is dishonest, incompetent, or insolvent. He may be an efficient businessman, a good man in his specialty, and a good manager. He may have a pleasing personality and a genuine appreciation of your problem and his duties. But the question remains—does this matter?

The critical situation arises with a subcontractor when the demands of the industry for workers, or occasionally the demands for materials, exceed the supply and the subcontractor is unable to meet his commitments. This is usually a situation which he cannot forecast, and over which he has no control. But this is the chance he takes; sometimes there are plenty of workers and he makes money; if there are few, he may lose money. A large firm can average out the jobs, but a small one may go broke as a result of one bad job. The subcontractor has no right to a profit—he must earn it.

It does not pay to force a subcontractor into bankruptcy for the simple reason that he has just so much money to put into your job, and if he cannot pay his debts on your job, your employer will have to pay them anyway. But the contractor can expect the subcontractor to complete the job on time to the extent of his financial ability. You are no more responsible for the subcontractor going broke than if he plays poker and loses his business; even less, since the subcontractor chooses his own game—his trade. If you are unwilling to push a subcontractor to the point that he will go broke, then don't recommend him in the first place—pick an established subcontractor who can survive a bad job. You may not be doing a small subcontractor a favor by giving him a job—you don't know if your job is going to be good for him or not. That's why you subcontract work in the first place.

In short, a superintendent can be sympathetic with the troubles of a subcontractor only to the extent that the superintendent will not ask the impossible just because it's in the contract. But the superintendent is obliged to ask the subcontractor to do everything possible to enable the general contractor to complete the work. This includes relinquishment of the contract.

9.8 ALLOWABLE CAUSES OF DELAY

In a subcontract, as in the general contract, certain events are cause for delay in the work, if so stated. The subcontract may refer to the general contract, in which case allowable delays are the same. The usual allowable causes are events beyond the control of the contractor, such as bad weather, storms, strikes, and material shortages. On the other hand, even the phrase "beyond the control of the contractor" is capable of wide interpretation. In one instance, materials were stored on a job site inside a prison and stolen by prisoners; the contractor was not allowed to have a guard on the job outside of working hours. The architect held that the theft was *not* "beyond the control of the contractor."

You must check the subcontractor's "allowable causes of delay" for each contract.

9.9 PERFORMANCE BONDS BY SUBCONTRACTORS

Some contractors rely on performance bonds from subcontractors of doubtful financial ability. This costs 1 percent more on the subcontract work and supposedly assures that the work will be performed. If the subcontractor fails to keep up, you can then call the bonding company to take over the contract.

By this procedure, the general contractor is buying insurance against loss and substituting the judgment of the bonding company for his own in determining the financial ability of the subcontractor. There are drawbacks to bonded subcontracts, as you must deal with a bonding company, well equipped with legal help and dedicated to the minimization of losses on the job, rather than with a subcontractor. The subcontractor must admit being unable to perform before the bonding company will give any relief; and any arrangement made with the subcontractor to continue the job by others, which does not relieve him of his responsibility, will involve the bonding company. Furthermore, bonding companies have simply refused to pay, leaving the general contractor no alternative but to sue for collection. Any arrangement that makes litigation necessary is not satisfactory. The simplest arrangement is to be sure you have not overpaid the subcontractor; if there is to be any suing done, let him do it. You can at least complete the job with available funds, while the matter works its way through the courts.

Arbitration clauses in a contract permit a great deal of flexibility, since matters are settled by construction people, and disputes can usually be settled while the job is still under way. An arbitration hearing is preferable to a court case; in arbitration, you can tell your own story, question those who challenge it, and establish the facts by discussion. In court, all evidence must be presented in a formalized fashion, questioning of other witnesses is through an attorney, and much information that you consider important may not be admissible evidence. Also, your attorney may not know what to ask you, and your opponents may be able to make unchallenged statements because of the difficulty an attorney has in understanding the whole situation—he isn't a contractor. In arbitration, you will be explaining to a contractor rather than to a judge and a jury of people unfamiliar with construction, and the rules of evidence do not apply—there are few "objections." However, it is important that construction professionals form a majority of the arbitrators.

9.10 GENERAL'S RESPONSIBILITY FOR SUBCONTRACTOR'S DEBTS

Usually, the general contractor is responsible for any debts the subcontractor incurs for the job. If the subcontractor's paychecks bounce, you may be called on to cover the checks or to cash them. You may also be called on to pay C.O.D. shipments for subcontractors. Usually, a subcontractor is refused any but C.O.D. shipments when his financial situation is very shaky or he is behind with payments to the seller. In one instance, however, a major glass company refused to ship a carload of plate glass to a contractor without C.O.D. payment, although the contractor had several million dollars available in current funds. Such situations arise because the buyer wants but a single shipment and has not established a credit rating with the seller. There is no reason to refuse payment for subcontractor's C.O.D. shipments as long as your employer has adequate funds; the amount is deducted from the subcontractor's next payment. Be sure you don't accept *custody* of the material, however; the subcontractor is responsible for breakage and for checking the shipment, and you don't want this responsibility if you can avoid it. If the subcontractor is financially shaky, you may have the material sent to your firm to be sure you have title to it. If the seller balks at selling to you (for some sellers will not sell to general contractors), he can generally be convinced that he is better off getting his money than preserving his trade customs.

This liability by the general contractor is due to lien laws on private work, and payment bonds on public work. It varies with different states and corporate organizations; consult your employer for applicable rule on your job.

9.11 BACKCHARGES

When the subcontract clearly states the duties of the subcontractor and the extent of your authority over him, your job is much easier. Subcontract clauses that give you virtually unlimited authority may be more of a hindrance than a help. For example, if the subcontract says that the subcontractor may store his material "and move it as directed by the superintendent," your order to move material may be challenged by the subcontractor on the ground that it is "unreasonable"; there is no explanation of what is reasonable. If the subcontract states, for example, that the subcontractor's material must be moved immediately on request if not stored where directed, but within 48 hours if stored where designated by the superintendent, there is no question as to what is "reasonable." Such a clause should allow the superintendent to move material at the subcontractor's expense, especially if the subcontractor is off the job.

Labor for cleanup should be agreed on before the subcontractor starts work; the subcontract cannot anticipate conditions. Often, the subcontractor will accept backcharges, as the superintendent has laborers available for the work while the subcontractor has none. On large jobs, a small power cleaner—this is like a street sweeper but will pass through doors—saves sweeping cost, but if the subcontractor is to pay his share of the cost of operation of a sweeper, it must be by prior arrangement. Some subcontractors, particularly the plasterer and terrazzo firms, will make stains on the work of others which cannot possibly be removed; backcharges are insufficient to ensure their compliance with cleanup orders. They must be stopped from working or removed from the job when their workers become sloppy.

Often charges between subcontractors are handled through the superintendent rather than directly. For example, the mechanical contractor may require some electrical work, but the electrician does not want to collect from firms other than the general contractor. The mechanical contractor gives you an order for the work, and you order it from the electrical contractor. Since you can deduct it from the amount due the mechanical contractor, there is no doubt about your ability to collect; the electrician would have no such recourse.

On every job there is a certain amount of trading of extra work, and if controlled, this cuts down paperwork. The carpenter builds a toolbox for the electrician, and the electrician does some temporary wiring or makes an extension cord for the carpenter. This throws off your job labor cost control, unless you keep a separate cost account for such work. The foremen involved usually know the approximate value of the exchanged work. In this example, you have no way to tell if the electrician is correct, even if a written order was used; if a foreman makes the deal, he usually will be careful that he is not making a bad deal. From the standpoint of the home office, such deals should be prohibited, as they can result in waste; from your standpoint, you must consider whether compliance with regulations designed to reduce waste are actually accomplishing the purpose. When regulations and common sense conflict, you are paid for judgment, not for compliance. If your judgment is poor, you will be replaced. The most rigid of organizations do not slavishly follow their regulations, as every military veteran knows.

If the subcontractor's foreman refuses to sign a backcharge slip for work you have done for him, attempt to get his signature on a description of the work and the circumstances under which it was done. The foreman may not want to obligate his company, and may not have the information at hand to tell if the work is his obligation or not. The description should be complete enough that the subcontractor's supervisor will recognize the work as his obligation and accept the charge.

9.12 SUBCONTRACTOR'S INSURANCE

The general contractor is responsible for workmen's compensation and liability insurance under the laws of most states for all employees on his job. Consequently, subcontractors are required to furnish certificates of such insurance (as well as for other insurance, according to the subcontract) before starting work. Sometimes these certificates are not available when you want the subcontractor to start work. Remember that these certificates may not be for the benefit of your employer but may be required for either the architect or your employer's insurance company; consequently, the contractor must obtain them. If a subcontractor is allowed to start work without furnishing these certificates, you may not be able to require him to furnish them later.

There are two instances when these certificates may be waived, and in both cases you should get the permission of the contractor to do so. If the certificate is really in the mail or in the process of preparation, the insuror may be contacted by telephone. If some types of insurance, particularly workmen's compensation, are not actually being secured by the subcontractor, the contractor may assume the obligation and pay the premium on his own policies. The subcontractor would usually be backcharged for this premium amount.

9.13 REASONS FOR SUBCONTRACTING

Eventually, you will be asked if work should be subcontracted or if you would prefer to do it with your own force. Initially, this decision can't be made on the basis of relative costs—no costs are available. When your own work force has completed work for which subcontract bids were available, you can decide on the basis of reported costs.

The decision to complete work previously subcontracted depends primarily on the ability of the persons who will direct the work. Can you, with the help of the people you know to be available, do as good a job as a subcontractor would? Do you know as much about the business as he does? To some extent, you are relieved from a difficult part of the business—bidding and contracting. You have a market. But unless you are able to supervise a foreman or trade superintendent as well as he does, you are under a severe disadvantage. The usual answer is that you will hire a foreman who *does* know—perhaps the foreman is already doing this work. But will this foreman work for you as he did for the subcontractor? If he does not, will you know the difference? How much supervision does he get from the subcontractor? A foreman can be very good when working for a person whose knowledge he respects, but can become very inefficient when he feels he has advanced a notch—and that there is no one to judge his ability.

You can acquire some of the ability to supervise other trades by careful attention to them while the work is sublet. You can learn just by watching, and by inquiry of the foremen. This requires you to be on the job constantly—and to take care of your paperwork outside of working hours, in many cases. Such effort does not show up on the first or second job, but eventually it makes you a *general* superintendent—not just a carpenter foreman who calls subcontractors.

If you are going to tackle other trades' work, you must be able to:

1. *Obtain good people.* The usual difficulty is that one job is too short to discriminate between workers or to attract good ones. On the other hand, if you have the largest job in the area no subcontractor is any better off—he'll have to hire new workers anyway.
2. *Buy materials at the same cost as does the subcontractor.* If your contractor has a proper license, you can usually do this, but he may have to set up a separate corporation for this purpose, particularly for mechanical work. On a large job, you can find a supplier, but local wholesalers on small jobs may want an excuse before they will sell to you. The corporation may exist only on paper—without even a bank account—but it serves as a trade buyer.
3. *Obtain a municipal, state, or county license.* In some areas, this may be very difficult, as license applications are approved by other subcontractors in the area. These people may be more inclined to license a firm apparently not connected with a general contractor.

In short, if you have employees with the requisite skills and you have enough work of the specialty to keep them at work, there is no reason to subcontract your work. There is nothing automatic about any savings you may make in this way. One way to acquire experience is to let work on a cost-plus and guaranteed maximum basis, so that you can supervise the subcontractor to some degree while retaining the value of his experience, and have access to otherwise confidential information about operations.

9.14 JOB PROGRESS MEETINGS

Some firms are firmly committed to weekly meetings of all supervisory personnel to discuss delayed work. The advantage of such meetings is that if one subcontractor passes blame to another, the second person is present to defend himself. On large jobs, these meetings may be recorded for reference and review by the home office. You may preside over these meetings—the senior person on the job should do so. If the contractor is on the job, he may prefer to preside.

If these meetings are to be held, there are some points that will make them of greater value:

1. They should be on the same day each week; Wednesday is a good day, since there is time in the week, both before and after the meeting, to get things ironed out. Also, Wednesday is the end of the *pay week* for many firms, so the status of work can be reconciled with labor-cost reports at this time.
2. Make up a list of items to be discussed. If you find that some subcontractors are apparently not involved, tell them so ahead of time—don't require them to sit through a meeting if they have no business there, or you will then find they are hard to locate for meetings at which they are really needed.
3. If only one or two subcontractors are involved on an item, try to settle it before or after the meeting. Don't take up the time of all the supervisors to discuss a matter that could be settled privately with one subcontractor.
4. Keep the discussion on the business at hand, and cut off recitals of back misdeeds and future general promises. Don't allow the meeting to degenerate into a discussion between two subcontractors who should settle their differences privately.
5. Let each subcontractor know ahead of time what will be brought up so that he'll know whom to send to the meeting. If material delivery is important, get the subcontractor's expediter or office engineer and let him know what will be required. If there is a disagreement about delivery of motor starters, for example, between mechanical and electrical subcontractors (and isn't there always!), see that they come to the meeting prepared with necessary information. And be sure *you're* prepared! Don't go into a meeting armed only with what your foreman told you; check the job yourself.
6. Speak directly and clearly. If the meeting is being recorded, there is a tendency for everyone to be backward about speaking out.

Speak as you would in a private discussion, and don't try to "doll up" your language. If you have nothing to say, stop talking.

7. Don't get into a discussion of excuses. You are after facts; at the meeting, it is not important *why* something hasn't *been* done—simply *what will* be done. You are dealing in future events. By cutting off accusations before they get started, you will not have to cut off counteraccusations and excuses. If one subcontractor dropped a motor on another's toe and broke both the motor and the toe, you're interested first in the broken motor and its replacement—not the cause of the broken toe!

9.15 JOB COMMUNICATIONS

You can usually avoid the necessity for progress meetings by prompt action on correspondence and by discussion with individual subcontractors. Remember, you can get a telephone hookup of as many people as you like at the same time (a conference hookup), either locally or long distance.

The subcontractors' foremen should have a mailbox at the same location as your own foremen's mailbox, and regular checks should be made to see that they pick up their mail at least twice daily. Dropping a dollar bill into each box occasionally will show who's picking up his mail promptly and make the boxes more interesting. You should carry a pocket notebook for memorandums, and if a subcontractor's foreman will use it, give him a notebook of the same size, perhaps a different-colored cover.

If you set up regular office hours, the subcontractor's foreman can find you; usually, he has no office hours and is hard to find. The hours just before and just after lunch are good times for you to remain in your office; if the foreman picks up a note at this time, he can see you immediately. If you are in regular contact with his office, he may ask you about his own deliveries, which will keep you in more frequent contact with him.

Individual radios are usually justified for foremen, with radios also in all vehicles. Set a CB channel for communication with all subcontractors and individual vehicles that have CB radios only, and encourage everyone to monitor this channel. It is often helpful to reach people off the job leaving or coming to work, or at lunch, by CB.

To reach foremen who do not have radios, you make a daily tour schedule of the job, setting for each foreman a time when you will be at a location near his work. In this way, the foreman doesn't have to look for you with questions that may be answered once a day.

9.16 RESPONSIBILITY OF SUBCONTRACTORS

You may assume that you know just what is meant by *subcontractor.* In general, *any firm that furnishes and installs material on the job is a subcontractor.* As a matter of legal definition, you may find that your specifications are broader, terming all suppliers of material fabricated for the job as *subcontractors.* Although these off-site subcontractors must comply with certain requirements for payment of their labor, you are rarely concerned with their labor force.

You may have subcontractors who sell material and then install it later, with contracts relieving them of the overall responsibility for their work. For example, a firm may sell you overhead doors or toilet partitions as a vendor; you must receive and store the material, and you have to check the shipment. The firm has a separate labor subcontract for installation; sometimes the same person will be doing business with you under two company names. He may be an agent for the manufacturer when selling doors, but he is an independent subcontractor when he installs them.

Under this arrangement, the installation subcontractor is not responsible if the material is damaged, short, or stolen. On the other hand, if he is unable to come when called to install the material, you may get another subcontractor to do it, or do the work with your own personnel. You have the material and may do as you like with it.

9.17 SUPERVISION OF SUBCONTRACTORS

Don't disregard work to be done by subcontractors because you believe the labor cost is none of your affair. The progress of the job is governed by the speed of subcontractors for most of the time a job is under construction, and you are therefore deeply concerned about progress. To be effective as a supervisor, you must be familiar with subcontractors' operations. One difference between a carpenter foreman and a superintendent is the superintendent's knowledge of the subcontractors' work.

Trade Jurisdiction

10.1 ORIGIN OF JURISDICTION

There is work on a modern construction job which requires workers with a high degree of skill, such as pipefitters, plumbers, electricians, sheet metal workers, and others. In each of these kinds of work (crafts), there are parts which require less skill. Under open-shop conditions, these less-skilled portions of the work can be done by lower-paid workers; some workers can do the less-skilled portions of many trades.

When U.S. construction trades began to become organized into labor unions in the late 1800s, one of their first claims was to demand equal pay for people in each craft, regardless of the degree of skill of the individual. In so doing, they established definitions of the work to be done by their craft, and of the work that might be done by lesser-paid workers or by members of other unions. As each craft union was formed, its charter defined the craft by the work it did. A large number of craft unions resulted, with constant bickering between them. Eventually, with the formation of the American Federation of Labor (AFL), these craft unions combined into trade unions, each of which included several crafts. Members of a modern trade union are sometimes interchangeable between crafts; at least the disputes are settled in the union itself. Examples of crafts in the carpenters' union, for example, are soft-floor layers, hard-floor layers, acoustic tile setters, piledrivers, and unspecialized carpenters, who cut wood and drive nails.

In modern times, *jurisdictional disputes* refers to disputes between these trade unions. However, probably much longer than trades or crafts of the present type have existed, work has been separated on every construction job so that workers most skilled in certain work have monopolized it, and these workers have resented work being done by those whom the more skilled workers feel to be unqualified. This occurred even when everyone knew his job was permanent; it even occurred among slaves, who were not paid at all.

To an outsider, jurisdictional disputes seem unimportant; but when workers are paid by the hour or day, and they know that their ability to keep working regularly depends on defense of their trade jurisdiction, it is to be expected they will be jealous of their privileges. It is unreasonable to expect that even the best-intentioned workers will disregard their own welfare in the interest of others. People may respect a no-strike agreement, but you cannot expect them to like it.

10.2 JURISDICTION OF SUBCONTRACTORS

The jurisdictional rules of the AFL give to the firm contracting disputed work the responsibility for assigning workers to it. In most cases, this gives the contractor the initial opportunity of assignment; by including certain work in a subcontract, he virtually assures that the work will be done by that subcontractor's trade. In fact, some unions claim work on the basis of whether it is subcontracted to a firm employing their members or not. If an electrician's contract includes the setting of lamp-post bolts along with the setting of the posts, the electricians may do the work; if the general contractor does this work, it will be done by carpenters. Similarly, a plumbing contractor will use plumbers on any kind of pipe he installs, although other subcontractors might use laborers, fitters, or even ironworkers. Subcontracts may require the subcontractor to perform all "incidental work of his trade," without any description of how his trade is to be determined. Even in open-shop work, the union jurisdictional determinations serve as the best-known authority for the determination of a trade. The subcontract may be more specific than described above, stating that the subcontractor is to perform all work of his trade "as determined by the National Joint Board" or "as determined according to the rules of the National Joint Board." There is quite a bit of difference between the two descriptions; National Joint Board determinations are limited, and disagreements usually arise on items that have not been decided. Consequently, the rules are important; they say, among other things, that the work is to be assigned by the contractor. There is therefore a line of circular reasoning; "the rules" may be such that the subcontractor must follow the decision of the general contractor!

Since the firm in charge of the work in question is required to make actual work assignments to a trade, assignment of work to a particular *subcontractor* by a general contractor is not, strictly speaking, a work assignment to a *trade*. However, since subcontractors recognize assignment to them as assignment to *their trade,* you are making trade assignments for all practical purposes when you decide which subcontract includes certain items. For example, on one job there were a large number of circular light fixtures to be installed in a plywood ceiling. The electricians were cutting these holes with a saber saw when the carpenters complained that this was their work. Since the electrical contractor had accepted this as part of his contract, electricians would do it, and the carpenters considered this a trade assignment—which, to them, it was. In this case, the superintendent's decision was that the work belonged to the electricians because they had to lay out the work, which was the greatest part of the item. However, if the holes were cut before the plywood was installed, the work was that of the carpenters.

When work obviously not of a subcontractor's trade is assigned to a subcontractor, it is not considered a trade decision. In many cases the electrical contractor will agree to do "all work as called for on the electrical drawings," and these drawings may include concrete bases and steel fences. For some such items he will hire workers of the appropriate craft.

10.3 JURISDICTIONAL DECISIONS

A dispute usually arises because the general contractor or a subcontractor has assigned personnel to a particular item of work, and the job steward for another trade believes it belongs to his trade. The steward will complain to the other trade's foreman or to the job superintendent initially, and they may recognize his claim—usually only if they are already aware that the work they are doing is not of their own trade. It is a source of surprise to outsiders that many union members will do any work assigned without comment—they feel it is up to every trade to protect its own jurisdiction.

If the steward or job superintendent does not comply with his request, the complaining steward will call his own BA, who will call the BA of the other union, and they will attempt to get together on the job and agree on who is to do the work or to argue the matter before the job superintendent. The BA who loses will often threaten to call out his members, although such threats are more common than are strikes. The work continues as assigned by the subcontractor responsible for the work—or as assigned by the general contractor, who will often delegate this decision to you.

The respective unions may then appeal to their own international (national headquarters) representatives, who will attempt to settle the matter; and if neither is willing to concede the point or drop the claim, to the National Joint Board in Washington. This process can be very fast or very slow. In most cases, the superintendent's decision will stand, particularly if the superintendent makes a point to complete the disputed item as soon as possible.

Contractors who have no labor contracts are not bound by the Joint Board decisions, but if they employ union labor they are in practically the same situation as those who have contracts. If a union goes on strike to enforce jurisdictional *claims,* the strike is *always* a wildcat strike, and other unions are forbidden to support the striking union. It is the policy of the American Federation of Labor that unions are to ignore picket lines posted to support jurisdictional claims, but it is also well known that union members will not cross picket lines. The refusal to cross picket lines, even when known to be unjustified, is one of the firmest union rules and will be followed when other working rules are disregarded. This is probably not due so much to a reverence for the rules as it is to the fact that one cannot cross a picket line without public exposure. For this reason, the use of entrances other than the one being picketed (which is sometimes required for legal reasons) is often welcomed by the workers. The National Joint Board is at times not active. The National Labor Relations Board, a U.S. government agency, has at times made jurisdictional decisions as well.

10.4 SUPERINTENDENT'S DECISIONS

As a superintendent, you have several interests in the assignment of work. You have a short-range interest in giving as much work to subcontractors (without additional payment) as possible, and you have also a long-range interest in completing all work as cheaply and efficiently as possible.

When a subcontractor must pay for work if it is assigned to him, the superintendent is tempted to give the subcontractor as much work as possible. The electrician on one job agreed to all the work of his trade, and a plastic ceiling under fluorescent light fixtures was claimed by the electricians. The electrical contractor was surprised to find that he was obliged to install, at no additional payment, a structure that looked to him like a lather's job–a suspended steel framework ceiling–which was not shown on the electrical drawings and could have been found only by studying all the architectural details. In this instance, there was no argument from other trades, and the superintendent's view was that the electrician should be familiar with the jurisdictional claims of his own union; general contractors do not negotiate with electrical unions, and therefore the general cannot be responsible for what the electrical contractor's association agrees to put into a contract.

Other jurisdictional decisions may be an invitation to the superintendent to pass work on to the subcontractor. The example previously given of light fixture holes in a plywood ceiling would be an added expense to the general if awarded to the carpenters.

The superintendent should never use his power in jurisdictional matters to attempt to require additional work from the subcontractor. Likewise, he should not make decisions in order to do the work with his own forces and avoid the subcontractor's markup. This gain is a short-range one; it will soon become evident what the superintendent is doing, and his employer will get higher bids.

You have a long-range and legitimate interest in reducing construction costs as much as possible; this is the object of your career. Consequently, your decisions should reflect award of work to the trade which is best equipped by training to do the work, and to the trade that can do the work at least labor cost. But this is often not possible; if local custom or other circumstances are such that your award on such grounds will not be supported by union officials or by the Joint Board, there is nothing to be gained by such an award. You would like to award work in accordance with trade practice—that is, to the person whose tools and methods are best adapted to the work in dispute, but there are other factors governing the award which may be more important.

10.5 SOURCES OF JURISDICTIONAL DECISIONS

The publications most commonly available are the *Green Book*, the *Blue Book*, and the AGC *National Jurisdictional Agreements* (also called the *Gray Book*). These may be obtained from sources listed in the bibliography at the end of this book.

The *Green Book* is actually titled *Plan for Settling Jurisdictional Disputes Nationally and Locally* and contains "Agreements of Record." These agreements, which date back for as much as 60 years, are recognized by the AFL as the basis for assigning work and are obligatory for contractors and unions. This book is well known to building tradesmen. The *Blue Book* is titled *Procedural Rules and Regulations of the National Joint Board* and contains the rules governing appeals by unions and contractors, requests for assignment of work, and the handling of wildcat strikes. This pamphlet is less well known, as it contains no detailed information of actual awards. The AGC *National Jurisdictional Agreements* contains both final and tentative agreements, and is a supplement to the *Green Book*. The *Green Book* and the *National Jurisdictional Agreements* do not repeat the same agreements. There are other publications containing union agreements, but they are not so readily available and must be obtained from business agents, personnel managers, and others who attempt to keep a complete file. The *National Joint Board* does not regularly print all decisions for general distribution.

Many of the agreements in the AGC booklet are between only two unions; other trades are not prevented from claiming the work. These are "not-attested" agreements and are so marked. Consequently, you must conform to such agreements only in regard to the unions that signed the agreement.

Unions often complain to the National Board that local trade practices are ignored in decisions, but this will not often be a concern to you; if the local trade practice is established, the business agent of the union presently doing the work will tell you about it. You cannot always rely on either getting any information from BAs or getting correct information. For example, on one job the carpenters' BA claimed the hanging of steel main tees for lay-in acoustical panels, which had been assigned to the lathers. The carpenters claimed they had done this work locally but did not mention that the work done was direct nailing of the same materials to wood joists—making the operation essentially the same as installation of a spline system. In this instance the work was given to the lathers on the grounds that hanging of continuous members by wires was a lathers' skill, for which carpenters were not prepared and for which they were not equipped with tools.

It is important that you include in your decision all available references; the union representatives should have a file of them, and your employer may have some of them. It is hardly to be expected that you would be able to make a decision that would be accepted, as written, at the national level. If you write it up well enough to convince the complaining BA that the other union has a reasonable claim, further trouble may be eliminated.

10.6 CAUSES OF DISPUTES

The trade unions have high respect for history, as evidenced by the trade training in the apprenticeship programs and by the continued emphasis on historical items in their trade journals. But although they make no attempt to change former decisions, new materials and methods fit poorly into the ancient trade framework. When a new method or material is introduced, the assignment of this installation to a trade may depend on:

1. The material used.
2. The function of the new material—that is, what does it replace?
3. The method of installation—what is the trade method used, and what kind of tools are necessary?

If any *one* of these methods of making assignments were consistently of carpenters is best defined by what they don't do; that is, the categories of work they have relinquished to other trades. As new work comes up, they can be expected to claim anything that is not clearly excepted. Many of the exceptions mean little when other trades are not on the job to claim their work. For example, screeds and one-board forms are the work of the finishers. Cement finishers commonly come to the job when the concrete is to be poured; screeds and side forms are already in place, as the carpenters have built them.

Some of the items to which carpenters have relinquished claim are:

1. Canvas, muslin, or similar wall material, to the painters.
2. Metal cove base and window stool, to sheet metal workers, but many items of sheet metal trim have been retained.
3. Wood lath, gypsum lath, and corner beads, to lathers.
4. All electrical and mechanical systems, to the mechanical trades, but there are continued disputes on structural parts of systems and of placing certain equipment. The sheet metal workers do not have unlimited jurisdiction to all ductwork, only to that of sheet metal and fiber.
5. Toilet partitions, to sheet metal workers.
6. All work of the trowel trades, for which a trowel is actually used, but there may be disagreements over incidental work. Supporting arches for bricklayers, for example, would be set by carpenters.
7. All thin finishes applied as decoration, to the painters.
8. Structural steel frames and structural sheet metal, to ironworkers and sheet metal workers, with various exceptions.

In general, carpenters may claim any work which requires tools that cut the work on the job or requires vertical alignment or layout on the job. Most trades protect their jurisdiction from the carpenters by increased efficiency, which is their reason for being separate trades in the first place. A lather rarely resents carpenters who occasionally install lath —carpenters are not good enough at it to present a threat to his job. A carpenter makes his living largely by his versatility—he could be pushed out of any part of the work by people who do nothing else, but only in larger cities can the all-around carpenter be dispensed with—or promoted to superintendent. Yet, crafts are constantly being formed within the ranks of the carpenters, such as mill men, finish carpenters, drywall applicators, and form carpenters. There are older separations, such as floorlayers, millwrights, and piledrivers. The floorlayers are divided into wood and soft tile layers, as these crafts require different skills.

The *ironworkers* have undisputed possess of *Class A* structural steel, that is, structural steel frames of buildings rather than odds and ends of steel which are not attached to the frame. Lathers do their own iron work

followed, most disputes would be eliminated. However, each decision is usually not merely a dispute between trades but between the methods of assigning jurisdiction by material, use, or method.

10.7 ASSIGNMENT BY MATERIAL

If all materials were divided up among the trades, each jurisdiction could be clearly established. New materials could be assigned according to the old material they most closely resemble. All wood would belong to the carpenters, heavy steel to the iron workers, sheet to the sheet metal workers, concrete to the finishers, copper wire to the electricians, pipe to the plumbers, and so on. Such an arrangement by itself would soon eliminate the carpenters, as wood is being used less and less; ironworkers would find themselves without pipe railing, and cement finishers would be setting precast concrete. Such ridiculous results are apparent.

Where material is installed by an established trade practice, and both the material and the trade practice of a new material have belonged to the same trade, there is little trouble. But as a sheet metal-faced board, sawed by carpentry tools, replaces wood wallboard, you can expect both carpenters and sheet metal workers to claim it; and as old materials are installed by methods requiring less job labor, laborers are constantly claiming them. The types of material used by lathers and carpenters—both of whom install sheet metal shapes in acoustical work—are not clearly defined. There is no fundamental difference between plasterboard lath and drywall except the size of piece. Since carpenters (millwrights) install equipment, they dispute the claim of the material to be installed with pipefitters, boilermakers, plumbers, and electricians, all of whom also install equipment!

10.8 ASSIGNMENT BY USE

The best argument for assignment of work by the purpose it serves in the building is that in this way all tradesmen are kept busy. Methods and materials change much more rapidly than does the size of the labor force, and a change in material can wipe out a trade, particularly in one area, within a short time. If bricklayers, for example, are given all curtain walls in concrete buildings, regardless of the material used, there will always be work for bricklayers. If building all interior partitions and installing all doors and windows were assigned to the carpenters, regardless of material (this has been done to a considerable extent), the carpenters will always have work. Such a system requires a considerable retraining and versatility of tradesmen and contradicts the usual trade practices. The strongest and most successful advocates of this method of assignment are the mechanical trades. The electricians claim anything that carries electricity, whether it be sheets, rods, or wires, and whether it is iron, steel, aluminum, or copper. When they go beyond this and claim anything that aids in the lighting of the space (such as luminous ceilings), they are on more controversial ground. If a water system were designed to carry an electric current, you may be sure that the electricians would want to install it. Disputes arise when new kinds of construction combine a mechanical use with a structural or decorative use.

Plumbers also follow this reasoning. If a pipe carries water, or if any item is necessary to the functioning of liquid-carrying or gas-carrying systems, they will claim it. This is even extended to toilet paper holders, whose connection to functioning of the plumbing is certainly incidental!

10.9 ASSIGNMENT BY METHOD OF INSTALLATION

You would prefer all work to be installed as economically as possible. This is done by assigning work to the trade that can do it most cheaply and quickly. Such assignment is in frequent conflict not only with the preceding two principles but also with the trade union belief that the skill required for a trade should not be reduced. The skilled trades naturally resent the use of lower-paid workers to install work that was previously theirs, but which, through changes in manufacturing, has been accomplished in the factory rather than on the job.

The purpose of prefabrication is to reduce costs. In the long run, this will mean higher wages, but the differential between skilled and unskilled wages will be reduced. It may come about that unskilled laborers will be eliminated entirely; some contractors now work in such a way that laborers *must* have a high degree of skill. The classical "work gang" of laborers is disappearing—you may find construction jobs with hundreds of workers but nowhere will you see more than two laborers working together. Work gangs and "bank foreman" (foremen who stand on the bank and watch) persist largely in public work and in noncompetitive industries.

10.10 JURISDICTION OF VARIOUS TRADES

In this book, no differentiation is made between the terms "unions" and "crafts." It is assumed that within one union there will be no jurisdictional dispute between crafts—the ironworkers, for example, may be depended on to properly assign structural ironworkers and reinforcing ironworkers.

As the original tradesmen, *carpenters* point out that they are the builders, and all other trades are branches of carpentry. Even architecture

and structural engineering were once branches of carpentry, a architects, using the original word for "carpenter" to describe the fession, until recently did not allow their members to practice bu and the structural engineers, having adopted the word originally appli people who operate engines, now want to monopolize "engineer." Ex for the installation of electrical wiring and plumbing lines (which is b part of the work of electricians and plumbers), carpenters may claim a will often do, work of nearly any trade. Electricians and plumbers are pr tected more by licensing laws than by skill or custom, particularly in are where the carpenter may still be a true *builder,* completing all work i a building.

Carpenters consequently claim work under all three principles of material, use, and method. They establish lines and grades, throughout the building, for practically all trades. Anything that requires complicated measuring and computation of dimensions falls to their lot, and most operations requiring line and grade are theirs.

For this reason, superintendents are usually former carpenters, and carpenters are involved in labor disputes with all other trades. Since they once laid wood flooring, they lay asphalt tile (in most of the country); since they once laid wood shingles, they now lay asphalt shingles; since wood siding was theirs, they will continue with sheet metal siding. Wall finishes are a source of conflict with the painter as papers become thicker and other finishes thinner. From wood ceilings, they have gone to include acoustical ceilings; from formwork, to setting iron and steel in formwork; from wood piling, to steel piling.

Carpenters have other advantages in establishing their jurisdiction. They are always on the job to claim their work. Many superintendents feel a close kinship with their former trade and try to protect its jurisdiction, and there are usually plenty of carpenters available. Since they do such varied tasks, there are usually simple tasks for the less-skilled people, leaving the more skilled workers available. The other smaller unions cannot provide such a variety of workers. This is particularly important also because the carpenters are directly under the control of the general contractor; if you have a job that could be done by a good worker of any trade, you know you can find a reliable carpenter. The carpenters' union has a much more flexible policy toward admission of people, and interchanging them between locals, than do other trades; consequently, there are fewer shortages of carpenters.

Carpenters not only work freely across jurisdictional lines, but exercise what might be called *secondary jurisdiction*—they will claim work of other trades if the other trade is not on the job. For example, if sheet metal workers are not on the job, carpenters will claim installation of metal cabinets as against all other trades. They may claim installation of reinforcing steel in the absence of ironworkers. In short, the jurisdiction

when necessary for lathing; this is usually confined to standard 1½-inch channel, but occasionally they do other ironwork. Ironworkers set Q-panels, a cellular steel deck, and steel decking for flooring. Ten-gauge and lighter decking and metal work in general belongs to the sheet metal workers. Ironworkers set stairs and railing with standard railing joints or welding connections. Plumbers claim only railings that have threaded joints, now seldom used. If corrugated metal sheeting with simple side laps is set on steel frame, this is done by the ironworkers; on other types of frames, by sheet metal workers.

The ironworkers and *boilermakers* have a complicated division of work. Complete boilers and sections are set by ironworkers, but portions are set by boilermakers. The ironworkers install smokestacks in hotels and office buildings, but not other stacks, and they install portions of boiler installations. Portions of the elevator contract will be installed by ironworkers—the manually operated doors (but not automatic doors) with frames and trim, and escalator trusswork. They do not install the elevator tracks.

Metal windows heavier than 10-gauge are in the jurisdiction of the ironworkers, and they install large fire doors. On occasion, scaffolding contracts have been let to steel erection contractors, who used their own trade, although there appears to be no claim to this work by the ironworkers. The same trade installs subframes as installs the sash. In general, it appears to be the intent that carpenters are to install residential-type sash, and ironworkers are to install commercial sash; possibly this reflects the realities of the situation, as ironworkers are rarely available on residential jobs anyway.

Ironworkers are the rigging trade, claiming jurisdiction of erection and dismantling of derricks, and handling of material by crane or hoist. They split monorail and conveyor installation with the *millwrights,* the ironworkers installing the structural members. Because of their structural steel construction, antenna and high-tension electrical towers are the work of the ironworkers, and because of their rigging duties, they share setting of electrical equipment, when power equipment is used, with the electricians.

There has been considerable disagreement between the ironworkers and *glaziers* (a craft of the painter's union) over aluminum members of trim and door frames. Since the installation of steel reinforcement in aluminum door frames is often an optional matter with the glazing contractor, he may choose a construction that does not require both trades. In general, the division of work has been made according to manufacturer, recognizing that distributors selling glass and trim employ only glaziers. Consequently, glaziers install stock aluminum trim and metal store doors. Presently, the agreement forbids steel reinforcement in tubular aluminum doors installed by glaziers. Metal entrances, other than stock aluminum sections, are generally the work of the ironworkers; it is intended to include in this category all bronze, revolving, and prefabricated units. The jurisdiction of an item may be controlled to some extent by advertising it in Sweet's (*Architectural Catalog File,* New York: Sweet's Catalog Service, division of F.W. Dodge Corporation) catalog (which makes it stock) or by the choice of factory prefabrication or assembly in place.

The ironworker-glazier agreement made in 1961 is long, complicated, and leaves room for further disagreement as new products are introduced.

The boilermakers, primarily a shop union, do not frequently appear on building construction jobs. Job-assembled boilers and distilling units are in their jurisdiction, and it is sometimes hard to define just what is meant by "job-assembled." There has been extensive argument over which items are really parts of the boiler to be assembled on the job and which items are separate accessories and therefore installed by *pipefitters.* On jobs requiring boilermakers, there should be a prejob conference with the United Association (plumbers and fitters) and with other trades working in the boiler room (ironworkers and *insulation workers*). The technical questions involved in such installations are too complicated for an inexperienced supervisor. Since boilermakers normally do the work of all trades in the shop, their anticipated jurisdiction encroaches on several trades, and several subcontractors are often involved. In Puerto Rico, there is an AFL Boilermakers' local which includes *all* construction trades, as there are no other construction trade unions on the island.

The *asbestos workers* are seldom troublesome, probably because it is a small union and their members are usually too busy to claim work of others. Also, they are on most jobs but a short time. They do not claim any home insulation work. There have been difficulties over the installation of insulation on the inside of ductwork at the shop; this may be done by sheet metal workers at the shop under the fiction that it is acoustical, not insulating, material. Of course, it serves both acoustical and insulating purposes, and the determination is based on information that is available only in the head of the designing engineer!

The *bricklayers* (the term *brick mason* can easily by confused with *cement mason,* an entirely different trade, and is therefore less often used) have a history as long as that of the carpenters and have a fairly well established jurisdiction. *Tilesetters,* in the same union, involve other trades, and therefore have more disputes. The bricklayers claim anything that is set in mortar, and some items which are not. They caulk around doors and windows set in masonry, except for caulking with oakum, which is the work of the carpenters. Large precast concrete items not set in mortar and which require riggers are shared with the ironworkers. Cement masons generally belong to the same union as the plasterers, but there are two unions with four crafts, of which two crafts may carry a card in either union. The cement finishers and plasterers may belong to either of these organizations, known commonly as the bricklayers and the *plasterers.* The bricklayers claim structural glass installation when it is set in mortar, but not otherwise. Since this material is generally spotted with mortar, similar to marble, they often claim this work. Since masonry contractors have shown no inclination to install items manufactured by glass companies, structural glass is usually set by glaziers. Masonry contractors install glass block which requires mortar, although it is often furnished by the glazing contractor.

The *electricians* handle all work associated with wiring and light fixtures, including all handling of their own material; their only concession has been to the elevator constructors, who do their own electrical work. Electricians lay their own fiber conduit, but so far have not appeared to claim other work in connection with it, such as concrete work or laying of tile conduit. On one job, they claimed shallow excavation; a compromise was reached, providing that if a worker could clean out the trench a few inches by pushing the shovel in front of him, it was the work of the electricians; if he had to put his foot on the shovel and dig behind the direction of travel, as a laborer does, it was the work of the laborers! Even then, the laborer's excavation was done by a worker on the general contractor's payroll, to avoid possible conflict.

The *plumbers, fitters,* and *sprinkler fitters* (all members of the United Association) often have different locals but settle their differences among themselves. These people make continuous efforts to obtain jurisdiction of exterior pipe work. Water distribution piping is usually installed by municipal agencies who refuse to recognize unions, but when these lines are installed by the general contractor, there may be friction. Water lines serving buildings will be claimed by plumbers, unless they serve only sprinkler lines, in which case they may be claimed by fitters. In many localities, this work is not done by union help, and qualified union workers are therefore not available. Although by an agreement of 1941, laborers are to generally lay sewer and storm drainage piping outside of buildings, plumbing locals will occasionally claim this work. Plumbers have been awarded lateral sewers from the main sewer into dwellings, in that the work shall be done by or "under the supervision of" the plumbers inside the property lines. At times, plumbers have extended this to include all sewers inside property lines even when there are no dwellings at all, and when the main itself is within the property. The rule is probably an outgrowth of city ordinances which require sewers to be installed by a licensed plumber. Plumbers have also acquired installation of albarene sink tops and bathroom accessories. They are the lead-burning trade, regardless of the purpose of the lead. Albarene is a natural stone used for sink tops in laboratories, but it is too expensive to be used unless acid resistance is necessary.

The *elevator constructors* generally exercise their jurisdiction by location of the work, a method entirely different from the principles previously mentioned. This jurisdiction has been accepted, so that elevators and escalators are assembled by elevator constructors. In effect, they are engaged in a shop operation on a construction site. This trade is very stable—a typical job takes months, and there are only a few employers in the country. The workers are therefore highly skilled in the work and complete a variety of trades, such as placing tracks, equipment, installing motors, cables or hydraulic rams, concrete roof slabs, and wiring. The wiring on modern elevators is quite complex—the most complex wiring on many jobs. Their job organization is somewhat like shops in that their helpers are actually apprentices rather than laborers. In effect, the elevator constructor draws a line around the elevator shaft, and says to other trades, "You stay out there and I stay in here, and we'll have no argument." These people do not negotiate wages but are paid on a formula based on what other trades receive. They consequently profit from all raises granted but without the necessity of striking for wages themselves. The organization is national, like the sprinkler fitters—workers travel all over the country as jobs are available.

The *operating engineers* have few jurisdictional disputes. The Teamsters occasionally claim equipment to which operating engineers feel they are entitled, and there are occasional conflicts with other trades over equipment that is mounted on equipment operated by the operators.

Operators' claims to work are much more expensive—they are made against nobody at all. They just want you to hire more workers. It is not unusual to have twice as many operators on a job as are actually needed. This occurs through a series of make-work claims: two workers on cranes where one would do, operators on engines that are quite capable of running unattended for days and would be unimportant even if they stopped, claims of a half- or full-day's pay on jobs where 1 or 2 hours' work is required, hiring of foremen and master mechanics although these people do not actually direct the equipment. The Taft-Hartley Law forbids *featherbedding,* defined as *pay for work not performed.* The operators are very careful to perform work for their pay, and therefore can stay clear of the law, but the work performed is often not necessary. To remain bundled up in a warm car while an air compressor puffs 50 feet away is, technically, working.

The laborers operate some small equipment, particularly gasoline-powered compactors and concrete buggies. But if you take the concrete body off a concrete buggy and substitute a fork lift, it is claimed by the operators. Laborers operate compressed-air powered drills and grinders but not compressors. Many items of equipment would be useful on a construction job, but only for short periods of time, and not if it is necessary to keep a person watching them. This particularly applies to equipment powered with one-cylinder gasoline engines. Open-shop organizations therefore have an advantage in labor costs; since such equipment is often not used at all on union jobs, the union contractor does not know the extent of this disadvantage. This is not important on many small jobs, even union jobs, because the working rules of the operators are not observed.

10.11 SUMMARY

The foregoing gives only a rough idea of the relative scope of the various trades. If you have decisions to make, talk to the most experienced people you can find, and try to get as much written material on the decisions as possible. Investigate fully the local practice; although there may be ample reason to modify other rules to follow trade practices as opposed to material and use principles, there is no reason to change local practice—this establishes the skilled personnel you need and is therefore the cheapest way. When you use local practice, even if the amount previously installed is very small, you obtain to some degree better workers and you get the best cooperation of the locals—even if the national union representatives are unhappy. It is the local you must live with, and when you can find some way to fight on the same side once in a while, you can get better cooperation from them.

Site Planning

Site planning for the construction superintendent is the layout of available space for use during construction, as shown in Fig. 25. Planning includes the following items.

1. Building layout and layout of reference points.
2. Construction offices, storage buildings, and change rooms.
3. Open storage and subcontractor spaces.
4. Worker and visitor parking.
5. Designation of space and surfacing for movement of trucks and equipment for hauling materials and for working on building.
6. Theft protection.
7. Safety fencing to keep nonworkers off the site.
8. Temporary utilities.
9. Goodwill and guide signs.
10. Designation of temporary use of permanent structures.
11. Temporary plant locations.
12. Off-site space.
13. Cleanup and trash hauling arrangements.

11.1 BUILDING LAYOUT

The A/E must furnish two points of line for location and direction (which means you must keep the space between them clear for as long as it is

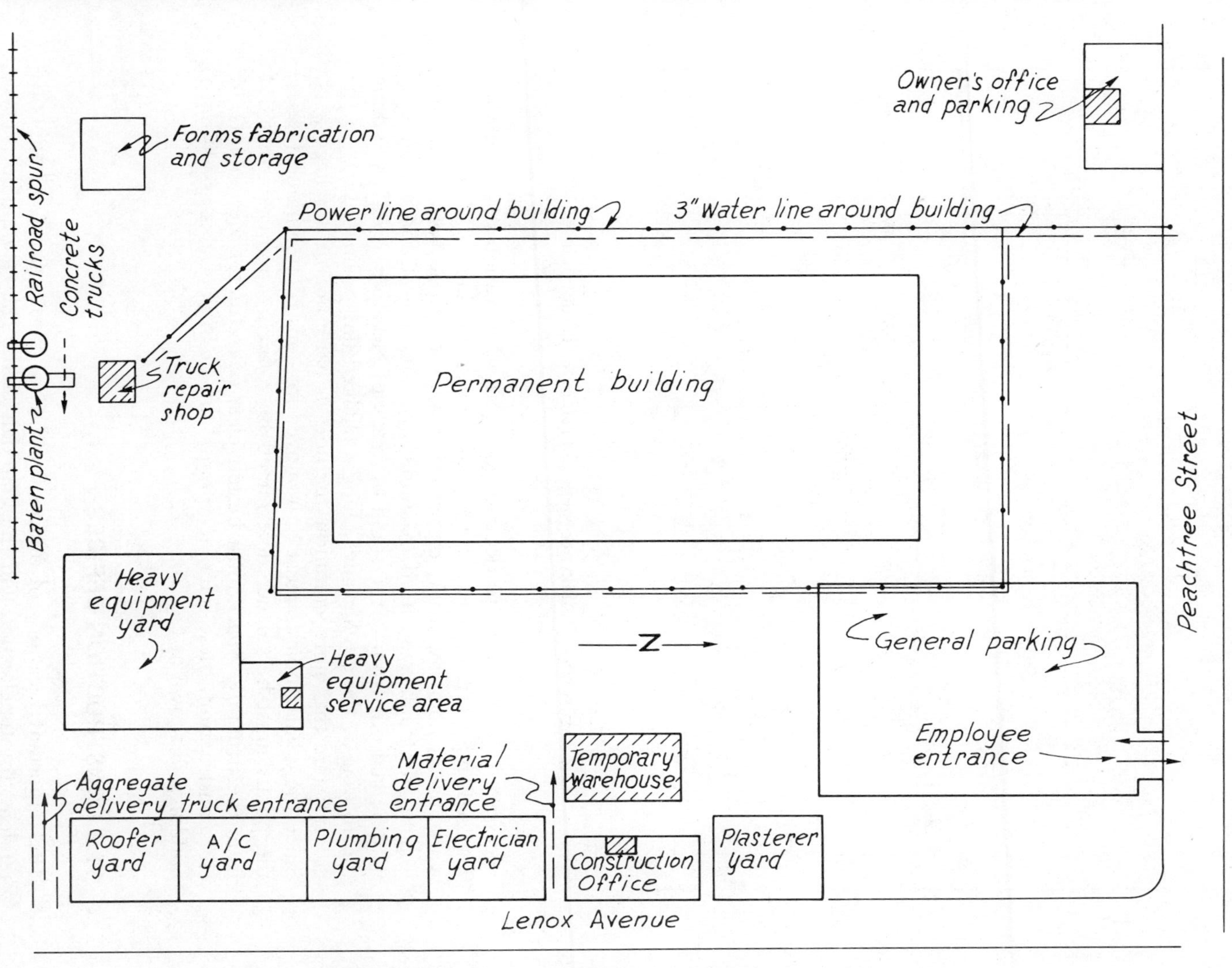

Figure 25 Construction site layout—Lenox Square, Atlanta, Ga.

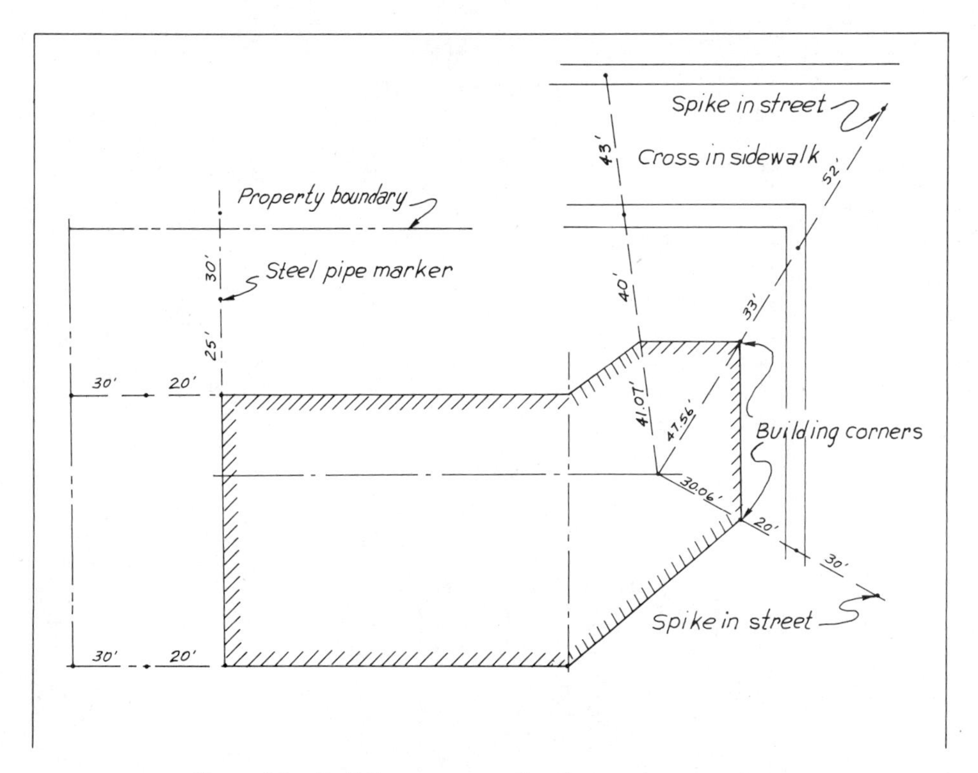

Figure 26 Building corners and reference layout points.

necessary to check your work) and one mark of elevation (bench mark). The A/E may furnish complete layouts and a continuous check on the work, as on highway work, or anything between these two. From this, you set the *points* you need. A simple, continuous wall footing building may be laid out from corners and no mark of elevation at all, by staking the footing as the ditcher goes along a chalked line or line laying on the ground; but the more common method is batterboards at all corners. Since these batterboards must be removed for construction in most cases, the points must be further determined by *reference* points, farther out from the structure. Space must be left clear of traffic for these points. In city construction, these reference points may be in sidewalks or adjoining buildings, as shown in Fig. 26.

11.2 CONSTRUCTION OFFICES

The important locations of offices for control are those for the superintendent, the timekeeper, and the materials clerk. They should be as close together as practical. The superintendent's office should be at a place where you, the superintendent, can observe as much of the work as possible, as well as the entering driveway. The timekeeper, if he is checking workers onto and off work, needs a location between employee parking and the work site. A small job or a very large one may not have a timekeeper of this type, depending on roving timekeepers or on foremen. If workers are checking in at the timekeeper's office, they should have a covered area to congregate. The materials clerk needs an office between the entering street and location of job and of materials storage. On some types of work, other materials clerks will be needed to check out stored material from open storage or from warehouses. All layout of buildings should be made with the objective of reducing lost time due to tool checkout, materials checkout, and time checking, as the amount of materials and time lost costs less than does delaying workers who must wait on material. If workers will actually start work in their respective work sites at starting time, checking onto the site is usually not justified. On a cramped site, all buildings may be on a second floor, overhanging the street (see Fig. 27). On a large job, none of the work may be visible from the superintendent's office. On one job, one of the owners erected a sentry tower, and, with a public address system, attempted to give orders throughout the job; needless to say, the workers did not long tolerate this type of centralized authority.

Since most site visitors will be looking for you, your office should not be too far from the gate, or you will have people and vehicles in dangerous places, interfering with work while looking for you.

Other buildings needed are determined by the job specifications, trade requirements, and requirements of the subcontractors. Separate entrances are sometimes used for various subcontractors, so that if one trade is being picketed, another does not have to cross the picket line. If you have old

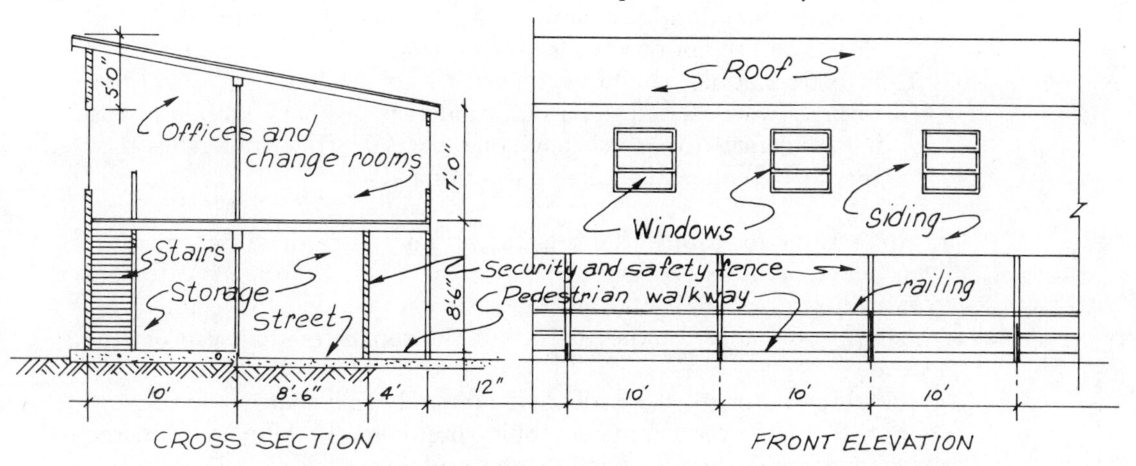

Figure 27 Urban temporary building.

buildings on the site, they may stay in place as long as possible so that you may use them for offices or storage.

You need change rooms for the various trades, for changing and storing clothing; some trades will not share change rooms with other trades, regardless of the number of people involved. You must furnish toolrooms for the craftsmen, to leave their tools in at night, and to store their tools not in use during the day. The contractor is often responsible for on-site theft loss of craftsmen's tools, and it is virtually impossible to tell if tools were actually stolen. It is helpful to mark the employees' tools with their names, as a service to them. It is good practice to replace tools when craftsmen have stated them to be lost or loaned to laborers (laborers normally use the contractor's tools), as the amount is negligible. Also, craftsmen usually have tools well beyond the requirements of their craft; if they are not reimbursed for tools lost, they will leave these tools at home. You'll then pay for them twice—once to buy them for use, and once in lost time to buy them and get them to where they are used.

Toilet facilities are provided in several ways:

1. By portable chemical toilets, single person, which may be moved by truck or crane, and emptied regularly. An important advantage is that these may be kept near the work force, with little time lost going back and forth.
2. By portable water closets, which may be set on top of sanitary manholes or connected to the underground sewer pipe, or in multistory buildings, to the sewer system.
3. By temporary toilet buildings, either built in place or prefabricated. This type is used with a concentrated work force on long jobs.
4. By early completion of permanent toilet facilities, with temporary closets and temporary walls, if necessary.
5. In offices, toilet facilities are built into the building, sometimes with showers as well. You may find that ordinary mobile homes for residence use are cheaper than special office trailers, as they are more frequently available secondhand.

In remote locations, the contractor may furnish either trailers for living units or hook-up facilities for them. Even in urban locations, the specifications or zoning authorities often allow one living unit on the job site, presumably for the guard, but the superintendent can occupy it instead.

You need storage for tools and materials that cannot be left in the open, and this may warrant temporary buildings. If these are too large to make of portable wood panels, metal buildings are used. These must be ordered well in advance of need.

11.3 OPEN STORAGE AND SUBCONTRACTOR SPACES

You may need fenced enclosures for material control for material that can be stored outdoors. Chain-link fences can often be rented from firms who install and remove them. The space needed also dictates the number of material clerks required, and your cost accounting methods affect the requirement for clerks. If your cost accounting requires that material be kept in a warehouse account until issued from the job, an additional building and clerk may be required for separated warehouse stock area.

Subcontractors are assigned space by the general contractor; each subcontractor is then responsible for his own fencing, buildings, and utilities. Their requirements both for space and utilities should be planned together. The subcontractors often call on the general and other subcontractors for work in their areas. The subcontractors offices and other buildings are normally scattered, as the subcontractors want the same material control as the general, so their offices remain near their material storage (as well as does their parking). By separate change rooms, toolshops, and material areas, workers who are in the wrong place and therefore who may be stealing, can be readily identified, since the subcontractor knows his own workers. If there are too many workers to identify by sight, different-colored helmets for each craft are helpful on the job. If a plumber is carrying a box of flooring tile, for example, it is most unlikely that he is lending a hand to a friend.

11.4 EMPLOYEE AND VISITOR PARKING

Employees will find their own parking, one way or another, but you are more likely to have satisfied workers if you make all possible arrangements in advance. Laborers, customarily, share cars and drive more people per car. Craftsmen have large boxes of tools; many of them keep their tools in their cars rather than use the toolhouse. The Internal Revenue Service has even recognized this hauling as a necessity, by allowing part of daily driving cost as an occupational expense. Consequently, the ratio of workers to cars is quite low in the carpenter trade but higher in the trowel trades, where craftsmen have fewer and lighter tools. It is desirable to allow craftsmen with tools to park as close to their work as security permits.

When the entire site is to be graded, a rotating area can be used for parking. Arrangements may be made with police to allow parking in areas where parking is not normally permitted, or with neighboring property owners to allow parking on their lots. You may need guards to direct traffic morning and evening.

11.5 MOVEMENT OF TRUCKS AND EQUIPMENT AROUND STRUCTURES

On cramped sites, dimensions of equipment to be used for erection and delivery shall be obtained and necessary clearances worked out. If several concrete trucks are to arrive on a city job, the location of the trucks' entrance should be worked out to avoid blocking street traffic and delivery trucks to your own job. If waiting spaces are available for concrete trucks off-site, they can be brought in as needed, ordered by their own radios.

11.6 PROTECTION AGAINST THEFT

Construction sites are exposed to theft more than any other type of facility. Protection is a *trade-off* between the cost of protection, particularly cost of guards, against cost of losses. Items stolen are those most readily sold, such as hand power tools, but professional thieves develop markets for major equipment, which is usually poorly protected with ignition or other locks. There is considerable casual theft by people who steal just because it's there, anything that looks valuable with no idea what it's for, such as custom faucet handles, or street light bulbs which will not operate except on special voltages. Equipment may be delivered to a job and signed for by someone who sees an opportunity for theft.

If you can persuade subcontractors to share the cost, guards become more economical. The mechanical trades—plumbing, electrical, and ventilating—have valuable equipment and will avoid storing equipment on the job. You will find that you have regular requests from these trades to pay for materials they are holding in their main warehouse. Such payments are usually optional with the A/E.

Protection against theft takes time, effort, and eternal vigilance, whether or not night guards have been used. Most theft occurs during working hours, by workers on the job. The constant traffic in toolboxes permits smuggling of small items off the job, and site security—that is, examination of materials and tools leaving the job—is very important. Much theft is never discovered, because one can't tell if the materials were actually delivered, or if the quantity needed was exactly what was ordered or delivered.

Authority for pickup of purchases and for making purchases should be carefully defined and followed. It seems harmless if you send a laborer to a hardware store for a box of chalk on the company account, but you may establish a precedent which allows *anyone* to buy *anything* from that store insofar as legal authority is concerned. If someone who is not even an employee charges tools to the company account, you cannot expect the seller to know he isn't authorized. Most sellers require a list of people and a sample of their signatures. If these people cannot leave the job, they should send a note with someone else. Be particularly careful to tell equipment vendors, repair people, and lessors (firms who rent equipment) whom they must see and how the person must be identified. Names on hard hats help somewhat in identification, as do different-colored hats for subcontractors and supervisors.

The Associated General Contractors of America (AGC) has published a well-written pamphlet that provides a security checklist, *Superintendent's Guide to Theft and Vandalism Prevention,** which is partially repeated, with drawings, in the following paragraphs.

The things you do can make a difference. While you may not be able to keep a "professional thief" from stealing your property, you can make it extremely discouraging for him, and totally impossible for the average hoodlum.

The layout and security planning that goes into a construction job site is often the difference between low losses and expensive thefts. A job site without guards, fencing, adequate lighting, or controlled exits makes a very easy target for even the most inexperienced thief.

There is no universally perfect program because job sites in different locations will require different protective procedures and devices.

Courtesy of Associated General Contractors of America

Solicit help from law enforcement and fire department officials. Well before you break ground or move in your equipment, you should meet with officials of the appropriate law enforcement agency (local police, state highway patrol, or county sheriff's office) and officers of the local fire department. Give them details of your construction, work schedule, project starting time, and the expected date of completion. Names of key personnel, with telephone numbers and places they can be located during nonworking hours are also essential. Keep the police and fire department

*See the bibliography at the end of the book for price and ordering information.

Courtesy of Associated General Contractors of America

posted on such things as delivery of critical material and unusual job-site activities that might require their special attention. Tell the police if and how your equipment is marked for identification.

Ask the local police to go over the site, and be sure to include them in all pre-job security planning. If you are going to use guards, it is sometimes good public relations to hire off-duty law enforcement personnel. Putting them on the payroll will increase their interest in keeping your job secure.

Neighbors can be a big help in watching your job. Neighbors and their children can become efficient watchdogs of your project if you

Courtesy of Associated General Contractors of America

solicit their help in a friendly way. Make it a point to personally contact neighbors in the immediate area. Emphasize your interest not in stopping crimes, but in the things you are doing to promote safety so that their children won't be tempted to play in the area and get hurt. Ask them to report to you any infringement of their yards and roads by parked cars, spilled dirt, dust, and avoidable noise. Bus drivers, cab drivers, and delivery route drivers passing your job site are also helpful; be sure to ask their help. Everyone is interested in your job; take time to answer their questions regarding why you are doing things a certain way, and if trees must be removed, explain why it is necessary.

Supervisory personnel should control keys. Key control is an essential element of access protection. Key issuance must always be based on actual need and not on convenience. The number of persons to whom responsibility of key control is given should be limited as much as possible, and an up-to-date log should be maintained listing the type of key issued, to whom, on what date, and for what purpose. All spare keys should be kept in a locked box or drawer. A sure method of making missing keys practically useless is occasional change of locks. Considering the value of material protected, this simple act is well worth the time and money. Security locks are available with combinations easily changed.

Lock or guard gates when not in use. Gates to your project should be kept to a minimum and all strange vehicles should be challenged. Whenever it is possible, guards should be provided during working hours to check vehicles in and out. When not in use, all gates should be locked.

Secure tools and equipment when not in use. Make sure that storage sheds or fenced areas are used to properly secure all tools and equipment. Keep cabs on all vehicles locked and ignition keys removed when not in use. Use metal shields on equipment windows where practical, and lock oil- and gas-tank caps. Disable machines with hidden ignition-cutout switches. Most losses are directly traced to carelessness by employees.

You must not overlook the most obvious opportunity for theft—tools in use, set aside for the moment on the floor or in an open truck. Theft of this type can be avoided only by constant vigilance and by company or foremen's names printed prominently on the equipment. Everyone must be on the alert for people whose hats or faces don't fit the tools they are carrying. The inspection at the outside gate may catch some of this type of theft. You and the guard—if there is one—should always notice people who are out of place—especially near the job boundary, where they may toss materials or tools to a place where they may be picked up later.

Your office should not be used as a toolroom. Even on a small job, the contractor and a party of owners or A/E people can fill and confuse your office, and a passing worker can pick up and carry away nearly anything without your noticing him—and the others don't know who he is. When you leave your office with tools in it, even for a moment, it must be locked. Such tools as axes, shovels, and sledge hammers will have a high rate of loss and cannot be adequately protected, but valuable tools such as engineering instruments and diamond-studded saw blades should be locked up, and only used when a person has actual custody of them.

Marking equipment with permanent identification pays dividends. One of the first things that a thief does after he steals some of your equipment is remove the plate on which the manufacturer has listed the model and serial number. A good portion of this equipment, once recovered, is never traced to the rightful owner. Even if you are absolutely certain that the machine belongs to you, positive proof is a legal requirement if you hope to reclaim it. An unusual alteration, mark, or attachment is enough, but the surest system is reliance on the original serial number. It is the most acceptable number in the FBI's National Crime Information Center computer system. Use a hardened steel punch (or etching tools) to duplicate the serial numbers in at least two places—one obvious and one hidden. Record the locations and the numbers. Then, if your equipment is stolen and recovered in another town or state, police can trace ownership back to you and you can make positive identification. Likewise, all company tools should be marked and provision made for employees to mark their own tools.

Make lighting work for you and against the thief. Lighting is very important to discourage theft and vandalism, particularly casual or impulse crime. Prior to job startup, temporary lighting should be in place, and plans should be made to expand its coverage as the job progresses. Consider renting lighting systems if company-owned systems are not available. Points such as the office trailer, material storage yard, and equipment storage area should be illuminated. Critical areas should be visible from the most heavily traveled road bordering the site. A good lighting system can also be spotted during the daytime by would-be thieves who might be casing the area. The small cost of overnight lighting is a real public relations bargain, as it tells law enforcement agencies that you want to help them protect your property, and cut down crime in the job-site area. It also shows dangerous excavations and discourages people who might walk in the dark, not knowing there is a construction site there.

Alarm systems do a better job. Alarm systems are becoming popular on construction job sites to protect supply trailers, portable buildings, and supplies. Some systems are highly sophisticated and the cost of installation and service may be prohibitive. However, moderately priced portable systems are available that will detect motion, activate lights, and sound alarms. This equipment can also activate phone calls to the contractor, security patrols, or police with a pre-recorded message.

Before making such installations, you should contact the police or sheriff's office, as they may prohibit installation of systems that make direct calls to them. The use of a private security patrol system may be necessary in such instances. Thoroughly investigate the qualifications of any private security company before hiring them. Check their bonding and insurance liability.

Encourage security suggestions from your employees. Employees can play a vital part in reducing losses of small tools and materials by constantly watching your job site. In preventing vandalism and theft, they can be working with you as well as working for you. Contrary to the belief of most employees, the insurance carried by the company does not cover *pilferage* of tools and equipment. Most insurance policies carry a deductible for coverage so that losses in any one instance would have to be $500 or, in some policies, over $1000, to be covered.

Thus, a lot of small day-to-day losses add up and must be paid for by the contractors. Don't be afraid to let the workers under you know that if they are caught stealing, they will be fired and prosecuted. Most labor contracts contain a clause that gives dishonesty as one of the just causes for which you may fire an employee. Prosecute those who steal, to let other employees know that you mean business.

Courtesy of Associated General Contractors of America

Signs discourage theft and vandalism. No trespassing, reward, and alarm signs not only discourage entry to your job site, but anyone caught trespassing on posted property can be prosecuted. Such signs and other warnings of danger can, in addition, protect your company from liability for injuries occurring to strangers on your property.

All construction property, whether it is fenced or not, should have

Courtesy of Associated General Contractors of America

adequate signs with lettering large enough to be read at a distance. These should be posted at gates and also on all sides of the job-site property. Signs that give warning or information about the marking of equipment with identification numbers will deter thieves and discourage vandalism.

Report theft and vandalism promptly. No matter how small the loss from theft or the damage by vandals, report all incidents to law enforcement officials. The information you supply to them promptly may save your job from a repeat visit or discourage the vandals or thieves from striking other construction projects.

11.7 SAFETY FENCING

A construction project is dangerous and attractive to children, and the law requires that you make a reasonable effort to keep children from dangerous machinery, excavations, and heights. A *reasonable* effort depends on what may be seen from places children travel or play, the effort required of them to get into a dangerous situation, expense necessary to protect the site, and the number of children normally in the area. The contractor has liability insurance to protect him in such situations, but this should not make him any less careful.

If you have a rural site of several hundred acres, with no construction visible from the boundaries, it is reasonable to block roads or established paths. If public access to your site is by boat alone, signs should suffice. An urban site adjoining busy streets requires 8-foot board fences, which are very difficult to climb without ropes or ladders. You cannot be blamed if, as happened, a group of college students goes over a 12-foot-high block wall by throwing one end of a roll of reinforcing mesh over the top and using it as a ladder! An urban site without sidwalks may require a 4-foot-high wire mesh fence. Safety fences can always be climbed easily; you can only make it reasonably difficult to discourage the curious.

In areas near an elementary school, utility trenches are a special problem. Fencing is not only expensive but impedes work; you may have to resort to guards during the time children can be expected to be around, as well as warning school authorities of your operations.

Figure 27 (p. 181) illustrates a common arrangement of safety fencing, pedestrian bypass, and temporary buildings as a cross section of the front line on an urban site with little space.

11.8 TEMPORARY UTILITIES

Temporary utilities for construction use nearly always include water and electric power. You may also have to provide compressed air and inflammable gas.

The water connection is made as early as possible. Urban jobs require only a connection to the public main and piping sufficient to reach throughout building sites with hoses. All municipal and most temporary water sources are adequate for both concrete mixing and drinking; job specifications usually require drinking-water quality for concrete. This kind of specification is meaningless abroad, where unsanitary water can be used for concrete, but drinking water may contain too many dissolved salts to be used for concrete. In some areas, seawater is used for concrete, although many concrete installations in which concrete was cracked open by rusting steel bars are blamed on the use of seawater.

Where temporary wells are not sanitary, a double water supply may be used. The locally available water is furnished through faucets with large signs that the water is dangerous and not to be drunk. *Do not drink—dangerous* is more effective than *Water not suitable for human consumption.*

Power may be from public utility lines on temporary extensions, from job generators with a temporary distribution system, or with individual generators in small areas. Since the general contractor usually furnishes water and power to the subcontractors, power machinery to be used and locations should be planned for the whole job and *engineered,* if necessary. It is economical to install 220/110-volt wiring with three wires, even if all 110-volt equipment is to be used, as wire sizes may be reduced. However, three-phase motors should be avoided, if possible, as two more wires and completely different generators may be required. A three-phase generator may be used for single-phase equipment, but a single-phase generator may not be used for three-phase equipment. A three-phase generator must be three times (or more) as large as any single-phase *demand* (electrical load) or it will be overloaded.

11.9 GOODWILL AND GUIDE SIGNS

There is a story that an old New York contractor, "Subway Sam," once went to a bank to request a loan. While he discussed collateral, he pointed out to the banker a line of his trucks coming up the street, with his name on the sides. He got his loan. The trucks had been rented and painted for the occasion.

It is unlikely that a contractor has this in mind when he has signs painted on his vehicles; still, construction is one of the last industries where a business is still owned by individuals. The largest firm in the nation still carries the name of its founders, Brown and Root (although it has long since been absorbed by a conglomerate). In every city in America, "Joe Jones and Sons" is as important a firm as "Jonesville Construction Company."

You can expect that your employer likes to see the firm name, and every operator wants his own name on his cab. Signs cost little, but they show everyone where the job is and who is the contractor. Some A/E's even restrict the size of the contractor's name on his job office! You should be generous with signs—identify your office, the job, the carpenters' toolroom, especially the solid fence (through which you may thoughtfully cut holes marked "for sidewalk superintendents") in cities. You need guide signs and arrows on nearby roads, and you will enjoy a tolerance of signs on public rights-of-way that the police will not allow to others. An entrance is just a hole in the fence until it has a sign. If your foremen use their trucks on the job, they will appreciate the loan of magnetic signs for them; when working inside public institutions and private plants with extensive grounds and security organizations of their own, such signs may be required. In turn, you get free advertising and the public thinks this is a big company, "to have so many trucks."

But signs are not just for fun—they are also to keep people from following you all over the job for instructions. If new workers are to report to the timekeeper, put up a sign "New employees report here." It saves you a lot of time. When you are busy, your employer is losing money; you need to run the job smoothly with a minimum of personal direction.

11.10 DESIGNATION OF TEMPORARY USE OF PERMANENT STRUCTURES

The most difficult case of site planning is the high-rise building downtown, when you are without any storage space except inside the permanent structure under construction. In such cases, the planning of sequence work may depend on the freeing of some of the space for parking and storage, as well as on completion of the job as soon as possible. You need the lower floors, usually provided with ramps, for the space usually provided by a lot. You need the stairs as soon as possible, for access. You need elevators, even with a temporary cab, in order to reach the upper floors. You need the fire water system for fire protection, and the water pumps both for fire and utility water on the upper floors. You need the walls, in order to cut off cold winds and to keep people from falling from the floors, and the power system for temporary power. You need upper floors to hang scaffold from, to erect the outside walls of the building. You may have, in effect, two operations—one to get your construction fully under way, and a following schedule for completion of the work itself.

On jobs with ample sites, you still need to get the permanent storage areas covered as soon as possible, to use for temporary storage.

11.11 TEMPORARY PLANT LOCATIONS

Temporary plants include concrete batching plants, precast concrete plants, and subassembly operations. You may have reinforcing steel bar bending facilities or an asphalt plant. On heavy construction, temporary plants may take up more site space than all other requirements combined. You need to plan the times when plants will be needed, moved, and removed. These plants vary widely and often do not have to be on the site itself.

11.12 OFF-SITE SPACE

Many of the facilities noted above can be placed in locations other than the site itself if land is available. Parking and storage can be rented; storage may be miles from the site. In many cases, vendors cannot store your orders in their own warehouses for lack of space, and to assure availability of materials, you must store them—even in a city other than the one where your job is being built. This is particularly true of high-rise structures and of structural steel, where day-by-day operations depend on daily deliveries, often going directly into the building from the delivery truck. Don't rely on delivery promises; if you have to have materials on a certain date, you should be able to store them on an earlier date. Delivery dates of construction materials are just not dependable enough that you can rely on the dates given you, except approximately. Conditions vary, but you should be able to store 60 days' worth of materials (other than stock material available from local warehouses) to have any confidence that you will meet your schedule dates.

11.13 CLEANUP AND TRASH HAULING ARRANGEMENTS

As materials must be stored to assure delivery, outgoing trash must be stored or hauled daily. On small jobs, trash can be piled for weekly haul, but usually this is a daily task. Even a $2 or $3 million job requires a full-time cleanup and haul crew, which means a full-time truck. Subcontractors usually are provided with trash services if they leave their trash at specified locations. You need to establish trash pickup points, trash chutes if above the second floor, and a system to get the trash off the job. In some locations you may still burn trash, but in urban areas this is now usually prohibited. You may want a scrap iron storage pile also, if you have a salable quantity. The first people you approach are the local municipal trash haulers, either public or private, and if they will haul trash, establish pickup methods and container requirements. Some items, such as scrap formwork, can be given away, but this may cause delays.

11.14 SUMMARY

As construction of the building is planned, the use of the site for this construction must be also planned. Sometimes the site facilities and space requirements are such that the construction itself must be delayed to allow for the incidental temporary facilities. On heavy construction, these facilities are planned as part of the overall construction; on building jobs, very little thought may be given to temporary facilities. The superintendent must study the schedule given him (if any) and plan his own site facilities. Similarly, purchasing for these facilities is seldom done by the purchasing department until they get requests from the superintendent, so they must be informed of job conditions and what is necessary from them if you are not doing this purchasing yourself. The delivery dates on materials, which should be provided you, must be planned to provide storage. This author has seen such planning ignored, even in competent firms with adequate experience at the work.

Safety

Construction work requires cutting materials to fit, which requires on-the-spot cuts. It requires work under the ground and in the air, in pieces which require workers in these locations. It requires hand placement of heavy weights and the use of poisonous materials. Each day's work is a different task in a different place. As a result, injury rates to workers make construction one of the most dangerous industries.

A craftsman who cuts material will sooner or later perform every variation, some of which are injurious or deadly. Safety practices intend to reduce this toll. To be effective, safety must take precedence over efficiency and personal liberty. This is the safety problem. Safety rules can be enforced only when safety is given importance as a minimum requirement of the work, even ahead of workmanship. Your reward is the ability to always say that no one has died of injuries under your supervision, but don't expect anyone to ask you, or to get any other credit.

12.1 SAFETY AND SUPERVISION

You are responsible for preventing injuries by safe practices, just as you are responsible for workmanship and production by efficient practices. Explain to each subcontractor that although you do not direct his workers otherwise, you will stop unsafe work if a foreman is not in sight at the time to do so. His carelessness is an infraction of the contract, which may be irreparable through an injury. Of course, you cannot stop his workers on doubtful points, but most practices that you would stop, such as grinding or welding without eyeglasses or shield, are not doubtful.

Rarely does ignorance cause accidents (but there are important exceptions, as noted in Section 12.2); workers know what to do, but they lack safeguards, or wearing them is inconvenient.

Employees work safely when the superintendent and foremen insist on it. Safety engineers, posters, and meetings are intended to achieve this. Employees and union representatives should be encouraged to report dangerous work methods.

12.2 DANGEROUS TEMPORARY AND UNFINISHED STRUCTURES

Unlike dangerous personal work practices, dangerous structural practices are not apparent except to the expert observer. Construction history is full of accounts of such failures as:

1. Failure of concrete floors supporting formwork and new concrete construction above.
2. Failure of scaffolding secured by bolts placed into new concrete, and used for lifting equipment.
3. Failure of personnel elevators.
4. Failure of slab formwork supports.
5. Failure of scaffolding.
6. Failure of structures used as erection platforms.
7. Failure of cranes or other lifting equipment.

These are all failures of engineering; that is, the faults are difficult to judge from experience, and there are no easily understood rules that can be used to avoid danger. Also, failures often result in death, sometimes for scores of people at a time. A superintendent can never be sure he is right, and engineers prefer not to take responsibility if they can avoid it. Design of temporary structures by accepted standards for permanent structures is often unnecessarily expensive; and these topics are not thoroughly taught in engineering schools. Few engineers are called on to design temporary structures at any time. In some cases, testing to failure (basically, the most certain and economical method of design) is the only sure way to design the structures. When these failures occur, studies disclose that the structure was inadequate and that the party making the study is not to blame. This is not very helpful in avoiding future situations. Several causes of collapse are often found, but rarely is a specific recommendation made to show how it should be done. On one job, OSHA (The Federal Occupational Safety and Health Act) officials fined everyone associated with the work, although they were very vague about which rules were broken and why their regular inspections did not report a danger. In general, OSHA rules have little to say that is pertinent to structural collapse; on the Willow Island tragedy described later, OSHA did not mention any studies or conclusions by its own engineers, but relied on a consultant.

1. *Failure of concrete floors* is almost a regular occurrence. As high-rise concrete buildings go up, formwork for each new floor is carried on the floors, usually two or three below. A construction rate of a floor a week is not uncommon, which requires the new slab to be supported by slabs below of 1, 2, and 3 weeks of age, none of which have reached their full strength. In cold weather, concrete is especially slow to set; and apartment house floors are designed to carry only a load of 40 pounds per square foot, so three floors could carry one if all were fully cured and the load was evenly distributed. None of these conditions is true; if a slab is reshored using wood wedges, the load on a slab may be *greater* than the weight of the slab above.

The lower floors, under these conditions, have lower maximum loads than have the upper floors. This author suggests an engineer's approval and *supervision* of the formwork, in accordance with the following criteria, on apartment buildings and offices (warehouse designs are much stronger).

(a). All slabs poured during the previous 3 weeks should have at least 50 percent of the original shores in place (not removed and reset).

(b). The deflection in all slab floors, at a representative number of bays in each pour, should be checked before and after each pouring or form-stripping operation. If the deflection or rebound exceeds 1/360 of the span (or other amounts set by the engineer), work shall be delayed.

(c). Slab forms, if supported by shores (rather than by metal scaffold sections), should be designed to permit failure of any one shore without progressive failure of others.

The danger of slab failure is greatest when flat slabs are used; it is reduced somewhat when capitals (large tops) are used on columns. Slab failures are progressive; that is, if one floor collapses, the chance is very great that *all lower floors will fail. It has happened many times.*

2. *Failure of scaffolding fastened to new concrete* is not common, because few methods require scaffolding to be supported by such fastenings. However, the 1978 collapse of scaffolding and formwork on the Alleghany Power Willow Island cooling tower in West Virginia, which killed 51 workmen, emphasizes the importance of this type of failure. The Willow Island scaffolding was designed by the engineer department of the design-

build contractor, with prescribed erection procedures. The causes of the collapse are obscure, and unfortunately the ensuing legal actions of such calamities often cause all parties to protect their information. As a result, the experience is not used to prevent further injuries. It is believed that part of the cause was shock from breaking a wire rope supporting a power bucket, further described in paragraph 7 below. In this instance, the scaffolding itself was used to support the upper pulley of a concrete bucket conveyor cable, and it is thought the cable broke.

3. *Failure of elevator cables* is not under control of the superintendent, as any elevator must be designed to national standards by an engineer specialized in that field. Where possible, arrange for a permanent elevator to be installed, with a temporary cab if necessary, as soon as possible. Elevator motors will be three-phase, however, and may be at a higher voltage than 220, so special electrical arrangements must be made.

4. *Failure of slab formwork shoring* is usually not serious if the fall is one floor in elevation. Where supports are higher, the support layout should be checked by an engineer. Forming subcontractors use a standard design, and the larger firms have structural engineers in their organization. When shoring is longer than one floor in height, the particular drawing the foreman is using should have an engineer's signature. The same is true of manufacturer-designed scaffolding-type shoring, although this type of scaffolding is less likely to collapse.

5. *Failure of scaffolding* is not as common as it once was, as engineer-designed metal scaffolding is almost universally used, replacing job-designed wood scaffolding. You need to know the rated load for any metal scaffolding used for formwork or for bricklayers, as the loads are greater than light metal scaffolding will support.

6. *Failure of structures used for erection platforms* can be prevented only by proper engineering design and supervision. Failures that have occurred were usually caused by connections or details that were not properly specified or constructed.

7. *Failure of cranes or other lifting equipment* also occurs because the load limits are unknown or are not observed. Don't experiment with equipment while working—make test loadings under conditions that allow collapse with minimum danger to workers. Obtain and observe load and *boom-angle* data by the manufacturer, and add boom-angle indicators to cranes that lack them. One of the important causes of crane failure is *shock, vibration,* or *backlash,* which are variations of dynamic loading—that is, loads caused by movement of loads, not from actual weight, and which last only a second or less.

Dynamic loads can occur in several ways. A common dynamic loading is caused by a suspended load which falls because of a broken cable or fittings or is dropped by the operator to avoid falling. The boom and cab structure, which is loaded at the time of failure in one direction, backlashes to the opposite stop, which the cables and crane parts were not designed to withstand. An operator can drop his load rapidly enough that the boom will bounce, or backlash, and capsize, breaking the crane off the carrier unless the crane is designed so that the load line cannot be freely released.

Vibration of the boom and other parts can be caused by repeated small blows in *harmony* with the natural frequency of the boom. Every *elastic* structure (and metals are highly elastic) will vibrate when struck, as a tuning fork does. A violin can break a glass, people marching can wreck a bridge (so soldiers break step marching over a bridge). Large structures have much lower frequencies, usually too low to hear; an engine or piledriver can be in harmony with them. When such action occurs, the dangerous frequency is avoided by increasing or decreasing the frequency causing the vibration.

12.3 OSHA REGULATIONS

OSHA regulations are based on federal statute and are constantly being changed or modified by the federal agency responsible for them. The greater part refer to wood scaffolding, being substantially a copy of earlier private regulations, and refer to heavy construction specialties such as air pressure and helicopter work. At this writing (1978), these rules are emphasized almost to the exclusion of all others, and the OSHA administration has still refused to furnish job-site advisory services; that is, if an inspector comes on the job, it is for the purpose of citing variance of the rules and for assessing fines. That is, he will not come to the job at the contractor's request to give advice on corrective action required. However, a recent court decision (1978) required an OSHA inspector to obtain a search warrant before entering a job site unless the contractor (represented by the superintendent) gives him permission to do so. Many people feel that OSHA inspectors and administrators take a legalistic approach, unrelated to actual hazards and accident rates.

You should ask your employer if OSHA inspectors are to be given permission to enter the job and if you have any OSHA records to keep or to file. On most jobs, you need only send the workmen's compensation accident forms to your main office, and further OSHA forms and reports are completed there.

During normal inspections, your firm will be cited both for general and subcontractors' violations; few superintendents avoid them entirely. In this text, safe practices are stated without reference; practices required by OSHA, but often not followed, are noted as an OSHA requirement. This text mentions only the most common standards and is by no means complete. Some OSHA standards are vague; for example, hard hats are required "where there is a possible chance of head injury." There appears to be no exemption or mention for persons erecting safety equipment; it is quite possible that the exposure to injury to these workers could offset the reduction of injury to the workers actually erecting the structure. Various regulations are enforced very exactly—particularly dimensions—but do not materially affect safety. The rules in the following paragraphs are largely based on the *AGC Guide for Voluntary Compliance with OSHA*.

12.4 PERSONAL PROTECTIVE EQUIPMENT

Personal equipment issued to individual workers includes hard hats (helmets), ear covers or inserts, safety glasses, face shields, prescription glasses, dark glasses, respirators (including rescue breathing apparatus), lifelines, safety shoes, and gloves.

Hard hats are nearly always used for commercial construction but not entirely in homebuilding. Construction follows practice in other industries in that it is easier to require hard hats of everyone than to attempt to discriminate between people who need them and those who don't. Helmets are of plastic or metal, supported by a flexible headband so that the exterior hard material is not directly in contact with the head, as shown in Fig. 28. "Bump caps," similar to those worn by baseball players, are not

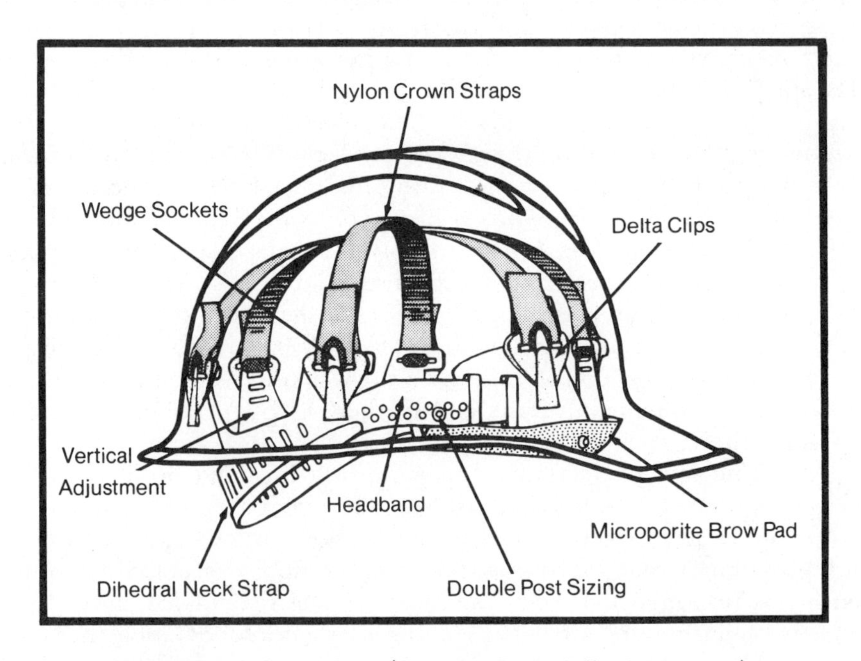

Figure 28 Safety helmet (Courtesy E. D. Bullard Company).

allowed on construction work by OSHA requirements. For winter work, cloth liners are used, and for people also needing face protection, helmets are fitted with face shields or welding helmets. Like all safety equipment, except prescription glasses and safety shoes, hard hats are furnished by the employer. Many workers wear their own hard hats, however, and some union contracts require members to furnish their own. Some people complain that helmets cause them headaches. Sometimes it is better to allow exceptions; this is a matter of judgment.

Ear protectors are required by OSHA in areas of high noise level, the level depending on the length of exposure. A sound meter is necessary to determine these levels, as is an individual recording meter with a master readout machine to determine the amount of noise exposure from the individual recorders. Some ear protectors are designed to allow low-frequency noises, such as voices, to penetrate but reduce high-frequency noises. They should be used by jackhammer operators and people working close to diesel engines. Individuals vary widely in how noise affects them.

Eye and face protection ranges from the worker's own glasses to welding helmets. Eye protection is needed for all chipping, grinding, or welding operations. Power cutting and driving nails, especially *hard* (concrete) nails, are also dangerous, but construction rules do not require glasses. A carpenter should wear safety glasses if he wears no other glasses, because his eyes are continually exposed to flying objects. Common nails being driven into wood can put out an eye, although safety standards do not recognize this hazard. Prescription glasses are available with safety-rated lens and frames, and can be equipped with side shields. Using them, a worker can avoid the necessity of wearing goggles to cover prescription glasses. You should do all you can to protect workers' eyes. Many commercial glasses in use are plastic and therefore unbreakable; you may identify them by weight, as they are much lighter than glass. OSHA rules prefer side shields for all required safety glasses but do not require them.

It is particularly dangerous to use a *burning torch* without eye protection, as the eye is exposed to both hot chips of metal and to *flash.* Flash is sunburn of the eye caused by exposure to ultraviolet radiation. The burn may be painless at the time, but the eye is irritated and painful for a day or more. One can get flash not only from an electric arc but also from the arc's reflection from walls.

Darkened eyepieces are available in a number of shades, to serve two basic purposes: a light glass, similar to dark sunglasses, for acetylene torch welding and burning, and a very dark glass for arc welding. Since an arc welder must push up the dark glass to inspect the weld, he also wears plain safety glass underneath for protection while chipping or brushing the weld. In hot weather, glasses and goggles become fogged by sweat, and workers resist wearing them.

Respirators, or breathing devices, are of many types, but standard construction varieties are:

1. Filter type to remove dust from air to be breathed.
2. Filter type, with gas-removal cartridge.
3. Independent-air-supply type, with tank or hose supplied air.

The *filter respirator* protects the wearer against dust, which is not a chemical poison, such as concrete grinding dust, blasting sand (fine sand used for high-air-pressure removal of paint and rust), asbestos dust, and dust caused by cutting brittle materials, such as brick and tile. These materials, except asbestos, are not dangerous in small amounts. Asbestos is much more dangerous; the lungs cannot rid themselves of it, so it accumulates over a lifetime and can cause cancer. If asbestos must be cut on the job, take it well outside buildings, do not allow anyone near the operator, and require the cutter to wear a respirator.

The *filter respirator with gas-removal cartridge* may be the same as the filter type, with an added cartridge. Certain gases are not dangerous in small amounts and may be removed by a cartridge designed for a particular gas. Some of these gases and mists are paints, pesticides, chlorine (in small amounts), and sulfur dioxide.

The *self-contained respirator* is similar to scuba-type underwater diving gear, and may be equipped with an air tank or air supply hose. This type is also known as a *rescue breathing apparatus,* used by firemen and rescue workers who must enter areas of unknown types of gases. A respirator is entirely independent of outside air and consequently may be used in any area, including those in which filter types are used.

A self-contained respirator must be used when entering tanks of leaded gasoline, which is very dangerous. It should be used with carbon monoxide and all other highly poisonous gases, including gas of unknown composition in sewers and deep holes. These respirators also furnish cool air, as they are supplied with air under pressure; like all gases, air cools when the pressure is released. Sand blasters use this type of respirator, not because the sand is dangerous, but because full eye and neck protection is needed, and this type of air supply is most comfortable.

Lifelines, lanyards, belts, and shock absorbers prevent injury to workers who lose their footing on high places. Current OSHA regulations require that workers more than 25 feet from a lower platform use some method of preventing falling, such as lines or nets. Since they must have a line attachment when moving from one place to another, the 6-foot line on each worker, the lanyard, is fastened to a long horizontal or vertical line along which the end fitting on the lanyard slides. For vertical lines, this fitting is made so that it may readily be moved by hand but will not move downward under stress. To reduce damage caused by the yank on the belt after falling, a shock absorber (a flexible piece of belting) may be installed in the lanyard, and a parachute type harness may be used instead of a belt. When lanyards are impractical, a net may be installed not over 25 feet below the working level.

Safety shoes are available in many styles such as boots, rubber boots, work shoes, or street-type shoes. They have protective concealed steel plates built into the shoe. Toe plates form a shell around the toe to prevent crushing by a dropped load, metatarsal plates covering the top of the foot just behind the toes spread any blow from this area to reduce severity, and a steel shank (plate) in the middle of the sole reduces the area that can be penetrated by stepping on a nail. The steel areas are limited, to allow the shoe to flex freely. Some safety shoes are available through retail shoe outlets, but the full line is sold by mail, principally to contractors. They are not required by OSHA rules, but you should provide an opportunity for your employees to buy them. Contractors should buy them for regular employees.

Gloves are essential for workers who handle rough materials, and they are furnished by the contractor for laborers and welders; laborers without gloves do less work.

12.5 JOB SERVICES

Job services affected by safety standards include:

1. Housekeeping.
2. Chutes for waste material.
3. Potable water.
4. Temporary lighting.

Good *housekeeping* requires that you collect trash continually or at least daily. You should remove nails as part of the form stripping. Since many subcontractors do not remove trash promptly, you must stay after them or, preferably, make a deal for them to accept agreed amounts for backcharges for trash removal. It is particularly important that when rolling scaffolds are to be used in a room, all small trash that might stop the rollers is removed promptly. OSHA rules allow employees to remain on rolling scaffolds during movement only if tools and materials are removed, which in practice is unlikely. You should have trash receptacles where fixed locations of workers, as at fixed saws and pipe threaders, make this practical.

Install *chutes* for throwing waste materials from floors above the first. These may be made of wood or from metal culvert pipe. They prevent material from striking people who might be below, and prevent materials from gliding or blowing far from the intended pile.

Potable *water* (for drinking) should be so marked, with a trash can for cups.

Toilets are needed at the OSHA rate of one seat and one urinal per 40 workers (for the exact number, consult the OSHA table), and their regulations do not allow for the fact that with the usual portable toilets, the urinal is unusable at the time the closet is in use, or vice versa. With a scattered work force, a larger number of portable toilets is required than the OSHA rule would require.

Medical facilities required are not defined by OSHA standards, other than that you should have a doctor-consultant to prescribe first-aid facilities. Some job specifications dictate a first-aid station, a nurse, or a doctor, depending on the size of the work force. You should take a first-aid course; these are offered by local Red Cross chapters. The military services give a similar course to all recruits.

Bibliography

This bibliography lists some of the easily understood and practical books currently available on construction, which may be useful for the job superintendent. You usually can rely on your employer for help in situations you do not understand, particularly in regard to extra claims, but you must recognize the situation that requires your employer's attention. In some organizations, you will be the technical expert for the firm and need references to terminology and procedures. The engineer, architectural, or construction college graduate does not have much training in construction methods, and you can readily acquire equal knowledge in this area with job experience and a reasonable amount of study.

This book list is very limited. No estimating books, no structural design books, and nothing about construction law is included. Most of your information will come from periodicals and manufacturers' publications, which are not included in this list. Few people have past issues of periodicals available, so there is no purpose in recommending magazine articles, and manufacturers' publications change each year. Many firms publish a pamphlet or handbook only once, revising it as soon as their supply is exhausted.

Books have a value that magazine articles and manufacturers' pamphlets do not; they are readily available for future reference when you need them. Also, they are generally more thorough and will give you basic information lacking in other sources; both books and pamphlets are necessary. And, particularly, don't overlook the plans and specifications you use every day—including the portions on subcontractors' work.

Architectural Drawing and Light Construction, Edward J. Muller, 1967, 450 pp., Prentice-Hall, Inc., Englewood Cliffs, N.J. 07632 ($17.95).

Written as a junior college textbook, this book is a quite complete and simply written explanation of architectural drafting as applied to homes. There are a number of practical details in regard to the use of drafting instruments, and sections on construction methods, electrical, and plumbing work. The book is practically self-teaching.

Architectural Graphic Standards, C. G. Ramsey and H. R. Sleeper, 6th ed., 727 pp., John Wiley & Sons, Inc., 605 Third Avenue, New York, N.Y. 10016 ($50.50).

A basic reference book for the architectural profession, construction supervisors, and students. This is a guide manual for architects, showing typical details of all sorts of work. It is advertised by the publishers as an aid for architects, engineers, decorators, builders, draftsmen, and students. The information in this book is furnished the superintendent on any complete set of plans he works from.

If you work on the kind of job (such as public work) where details of the work are clearly shown on the plans, this book is unnecessary. On the other hand, if you work with a developer who habitually uses very sketchy plans, this book is a source of many clarifying details. For example, how do you form concrete for an elevator door threshold? What is the height of school lockers? How is metal lath attached under all sorts of conditions? In many cases, items described in the specifications but not drawn on the plans, or items identified only by manufacturer, may be found here.

This is a collection of illustrations and specification descriptions. To the extent that you must make your own plans or work without them, it is a valuable reference.

Audel's Carpenters and Builders Guide, consisting of four volumes:

1. *Tools, Steel Square, Joinery,* 384 pp.
2. *Builders Math, Plans, Specifications,* 304 pp.
3. *Layouts, Foundations, Framing,* 304 pp.
4. *Millwork, Power Tools, Painting,* 368 pp.

Available from Theodore Audel & Company, 4300 West 62nd Street, Indianapolis, Indiana 46206 ($6.50 each volume or $24.50 per set).

The writer attempts to include in a single set of books all of carpentry and building for the journeyman and foreman. Unlike most similar books, it is assumed that the builder works without drawings and designs his job. It is written in plain language, with large print (although the book itself is small and a little hard to handle), ample illustrations, and has been recently rewritten to show modern construction practice.

The more necessary and useful portions of other carpentry books can usually be found in *Audel's.* No single book can be complete, and at times the writer attempts to say something about nearly everything. Consequently, many of the items are described too briefly to be of use. Do not expect, therefore, to find all you need to know about any item in *Audel's*—or in any other book. If you look for a specific point, you can often find it; if you want to know how to build a new kind of job, you will find the information you need scattered around in various places.

For the new superintendent who needs a reference for the carpentry trade, *Audel's* is highly recommended.

The Blue Book, the common name of the *Procedural Rules and Regulations of the National Joint Board,* for settlement of jurisdictional disputes in the building and construction industry, and of Appeals Board procedures. American Federation of Labor, 1965, 29 pp. Available from Associated General Contractors of America, 1957 E Street, N.W., Washington, D.C. 20006.

The superintendent who must make decisions on jurisdiction needs this book as a supplement to the *Green Book.* It is not as well known or as often used as is the *Green Book,* but it explains the procedures to be followed in settling disputes by the Joint Board. In this booklet, the contractor's and the union's responsibilities in the settlement of disputes are explained.

Building Construction, W. C. Huntington, 4th ed., 1975, 734 pp., John Wiley & Sons, Inc., 605 Third Avenue, New York, N.Y. 10016 ($19.50).

This standard textbook for architectural schools, first written in 1929, has been edited to remove descriptions of obsolete construction. All building products are described, from the raw materials to wall and finish details. Many terms are described, and the book is basically a guide to terminology.

Such information is of assistance to the job superintendent who wants to acquire a general knowledge of plans. This knowledge is chiefly useful as a matter of general education or to pass contractor-licensing examinations in which this type of book is used as a source for questions. Such examinations may be written by an examining authority who is an architect or who lacks material to write other types of examinations.

This book is not otherwise of great interest to the superintendent, as he usually has plans and specifications that are more detailed than descriptions here presented. Few methods of construction are described.

Building Estimator's Reference Book. Frank R. Walker Company, 5030 North Harlem Avenue, Chicago, Illinois, 60656 ($19.95).

This reference for methods and estimating has probably the greatest circulation, since 1915, of any book written for the construction industry and is used as a reference by practically every contractor in the country. Although termed a "reference" book, it is more readable as a methods text than are most more recent junior college textbooks in the subject, although the treatment is very thorough—which discourages many students.

The small page size so familiar to all, used for the first 18 editions, has finally been changed to a 5½- by 11-inch size, which is much easier to handle. The Construction Specifications Institute standards are used for numbering; these numbers are used in most architect's specifications.

The price is low as compared to other publications, and has not been raised as much as have other book prices.

Commercial Carpentry, apprentice curriculum material by Donald W. Diehl and Wayman R. Penner, 1974, Associated General Contractors of America, 1957 E Street, N.W., Washington, D.C. 20006. About 500 pages, in two loose-leaf volumes.

The new and current detailed instruction exercises for carpenter apprentice programs. This is a useful book for junior college instruction also, and as a reference by superintendents.

Construction Contracting, R. H. Clough, 3rd ed., 382 pp., John Wiley & Sons, Inc., 605 Third Avenue, New York, N.Y. 10016 ($19.00).

This book describes in a general way the running of a construction business, including organization (as a partnership or corporation), and explains the forms used, contract procedures, bonds, purchasing, and labor relations.

This is the most complete book of its type presently on the market: all types of construction businesses are described in a general way, without detailed consideration of any one, such as in the Deatherage series. It is a college textbook; sentences are long and words are often unfamiliar to tradesmen. It is therefore not easy reading. The writer is inclined to describe construction business practices in details of little interest to the job superintendent.

This book is a valuable addition to the library of the superintendent; it is worth reading, although you will find it hard going. There is no detailed treatment of paperwork, and construction methods are not included at all.

Construction and Professional Management, Rubey and Miller, 1966, 306 pp., University of Oklahoma Press, 1005 Asp Avenue, Norman Oklahoma 73069 ($15.25).

The writers describe applications or possible applications of management theory and statistics to construction management. As in many other books written by college professors, *professional* is used as a term for college-educated persons, and on this basis, "Professional management has been developed only in the last twenty years." It is implied that only young people educated in the subjects being presented can adequately manage the construction industry. Unfortunately, the training offered by universities in the construction field is very sparse, but it is pointed out that in the future there will be "less dependence on materials, money, and labor, and more management." This book presents an academic view of management, which is quite popular with some larger firms in other industries. Its application to construction is doubtful, particularly for persons without an engineering or statistics background.

The writers attempt to cover the entire field of management, equipment, human relations, estimating, and give a rather broad explanation of the various kinds of CPM and PERT techniques. The writers advise readers (even engineering graduates with statistics training) to seek the help of a statistician to understand portions of the presentation.

Computer applications are explained at length. Unfortunately, most of the source material and a great deal of the discussion does not apply to the construction industry. There is a rather complete bibliography with comments. Here, as in the rest of the book, the writers do not attempt to evaluate the usefulness of the subjects presented. The book is written as a textbook for a college engineering course.

CPM in Construction Management, J. J. O'Brien, 2nd ed., 1971, 321 pp., McGraw-Hill Book Company, 1221 Avenue of the Americas, New York, N.Y. 10020 ($24.00).

This publication presents the critical-path method in greater detail than do other publications on this subject. Some old hands will be surprised to hear that "prior to 1957, there was no disciplined method for planning and scheduling a construction project," but such statements are typical of CPM writers.

This book is quite complete and gives many examples of the various steps. A simplified version of manpower leveling appears with a statement that float time determines assignment of men by the superintendent. This is as close as any of the literature reviewed comes to the admission that the assignment of workers on the job is the actual end product of any system of planning or scheduling. Cost control with CPM is for some reason considered difficult. It appears that the writer is not familiar with budgeting and cost control systems in present use, which are very little affected by the application of CPM names to various items.

In general, the writer avoids unnecessarily complicated phrases and awkward professional language. The book is clear, well written, and probably the most complete work on CPM on the market.

Critical-Path Method, L. R. Shaffer, J. B. Ritter, and W. L. Meyer, 1965, 224 pp., McGraw-Hill Book Company, 1221 Avenue of the Americas, New York, N.Y. 10020 ($21.00).

Most writers on CPM have one thing in common—there is never any doubt that they know what they are selling but are not at all interested in any difficulties that may arise from its actual operation. Rarely will you find a presentation of any other planning or scheduling operation in an explanation of CPM. There is a tendency to create jobs that fit the outline being presented, and to ignore considerations that would not lend themselves to CPM methods of plotting.

This book is one of the best in the field, however. There are simple problems and solutions, which help a student who learns only from the book. The language is not unnecessarily difficult as compared with many other books in this area, although any analysis of CPM includes a number of terms that are rather vague to the newcomer—and possibly to the writer as well. A construction "operation," for example, is easy enough to define for theoretical purposes but hard to identify on the job. The authors realize this difficulty and do admit that judgment is required and that in the final analysis CPM is a guide for judgment, not a substitute for it. It would hardly be expected that an academic advocate of CPM would emphasize this attitude.

Design, E. E. Seelye, Vol. 1, 3rd ed., 1960, 700 pp., John Wiley & Sons, Inc., 605 Third Avenue, New York, N.Y. 10016 ($47.00).

This is an engineering data book, "the prime purpose of which was to

furnish an engineer with sufficient data so that he could design any civil engineering work without other reference books". This is quite a broad purpose, and the writer has attempted to give answers rather than to present theoretical discussions.

Since this is written for engineers, engineering education is assumed of the reader, and large parts are incomprehensible to a construction person without such a background. However, it is prepared for *any* engineer—not just for those who have learned the highly theoretical courses of the postwar years, and anyone with any study at all in engineering topics can understand most of it. There are portions easily understood by laymen; for example, you will find a typical swimming pool design, with required plumbing and filters; since engineering colleges do not include such topics at all, you are on a par with engineers. Illustrations of water distribution and other utility systems, and typical details, are likewise easily understood by anyone.

This book does for engineers what *Architectural Graphic Standards* does for the architect; it provides a simple and readily accessible source of information in standard details and standard construction. It is primarily a book of tables, although many of the tables refer to illustrations, and includes details such as elevation differences to lay out road crowns (or screeds), methods of testing, and allowable loads on columns of steel, concrete, and wood. For a person with some knowledge of structural design, it is very complete, and the person who intends to study structural design can profit from using this book as a guide during study. It is not intended as a textbook.

There are many actual designs shown, including tanks, manholes, concrete pavements, and retaining walls. Footing designs are shown for various loads, and it is comparatively simple to figure the load itself. An estimator may obtain data from this book to figure the cost of typical footings and other structural members shown in detail.

Formwork for Concrete, M. K. Hurd, 1963, 339 pp., American Co crete Institute, P.O. Box 19150, Redford Station, Detroit, Michig 48219 ($31.75).

It is characteristic of writing in the construction industry that book is published with considerable repetition of other books on the topic. Each writer assumes that superintendents will read his own and no other, and to some extent this may be true. This book illustrated, with large type and generally direct writing. This book result of the increasing concern by the American Concrete Institu the proper construction of formwork.

This is a "survey" book, in that a little bit of informatio about a great many different kinds of forms. There is too little

one item to be a great help to the superintendent, and unfortunately the book lacks both necessary detail and a bibliography of the sources. There are many photographs, usually of overall methods of forming rather than of details, and the book generally is of the quality of a well-organized collection of articles of formwork such as you might find in current trade magazines.

The ACI formwork standard is included as an appendix. As the only current standard for formwork, this should be available to every superintendent. The standard is available separately, however, for less than $1 from the ACI at the address above.

Of all the books available on concrete, only one—by Wynn and Manning—makes any mention of the most important problem: the time for stripping and backshoring of slabs in multistory construction. The *ACI Formwork Standard's* recommendations for this type of construction are not practical in many cases and are often not followed.

Formwork for Concrete Structures, R. L. Peurifoy, 2nd ed., 1976, 333 pp., McGraw-Hill Book Company, 1221 Avenue of the Americas, New York, N.Y. 10020 ($17.50).

This is the most recent and complete book on formwork published in this country. It is written in the style of the writer's other works, which are intended as college textbooks for first courses in construction topics.

This is a collection of available information on formwork, previously published chiefly by firms selling patent forms or ties of various kinds. The information has been organized and collected and engineering information has been added; for the most part, it is assumed that the reader has had a college course in structural engineering. The elements of wood design presented are rudimentary; a further study of structural design is required to use the formulas with any degree of safety. Tables on spacing of stringers, joists, and similar members are included.

This is neither a carpenter's book nor a detailed consideration of the cost involved in formwork design and construction. Although cost considerations are described, there is little information on comparative systems of forming, and there is little evidence that the writer has any experience with the construction described. Nevertheless, it is useful as the most complete collection of information available at present.

The *Gray Book* (the common name of the National Jurisdictional Agreements not printed in the *Green Book*), 1963, 154 pp., printed by and available from Associated General Contractors of America, 1957 E Street, N.W., Washington, D.C. 20006.

This book includes a number of decisions, agreements, and understandings on trade jurisdiction and is a valuable supplement to the *Green Book* (following).

Green Book, common name given to the *Plan for Settling Jurisdictional Disputes Nationally and Locally,* National Board for Jurisdictional Awards, and *Agreements and Decisions Rendered Affecting the Building Industry,* 1965, 147 pp., available from various trade unions and the AGC, including *The Lather,* 6530 New Hampshire Avenue, Takoma Park, Maryland 20012.

This is the latest issue of the official jurisdictional awards, which has been printed at intervals for several decades. These agreements are those which are approved as applicable to all unions. Decisions in the booklet are not classified either by trade or by date, making it difficult to read. This is a standard reference of all business agents and of most mechanics, so the superintendent on any union job must have a copy available.

Handbook of Heavy Construction, F. W. Stubbs, ed., 1971, 1440 pp., McGraw-Hill Book Company, 1221 Avenue of the Americas, New York, N.Y. 10020 ($54.50).

An important reference book for superintendents of heavy construction work. Each chapter of this handbook was prepared by a different writer, each writer an authority in his field. Construction methods in excavation, concrete work, steel erection and welding, timber construction, bituminous pavements, pipelines, foundations, and other topics are included. Many of the writers are supervisors of construction firms engaged in the particular specialty. Chapters written by engineers have more information about construction operations than do most books written for building construction.

The book is useful for the job superintendent and, incidentally, for the designing engineer. Some chapters, particularly those on welding, are more technically detailed than is necessary for the superintendent. This is not a simple book; the language is that of the specialists writing the chapters, but where they use uncommon terms, it is because these terms are necessary, not because they are dressing up the book with long words. The book is therefore worth studying if you are to supervise the operations described. For the building construction superintendent, much of the information is not necessary; but as you supervise larger jobs, you will use larger portions of the book. There is hardly a chapter that does not have some application, at least occasionally, to building construction.

This is not a "how-to-do-it" book, and portions of it are rather general, as might be expected with two dozen different authors. It is an important addition to the superintendent's library and one of the leading half-dozen construction books presently in print.

Lefax technical handbooks, tables, and blank forms. Catalog available from Lefax, 2667 East Alleghany Avenue, Philadelphia, Pennsylvania 19134.

Many of these pages of data, which are sold in chapters, books, or individual pages for a field notebook, are useful to the superintendent. There are several hundred kinds of forms. This firm also prints forms to order on field notebook size paper.

Minimum Property Standards for one and two living units, FHA No. 300, regularly revised, about 315 pp., furnished punched for three-ring notebook, unbound, from Superintendent of Documents, Washington, D.C. 20402 (on subscription, $10.25).

There are hundreds of government publications on construction, published at various times and by various agencies. Some of them have been duplicated by different agencies, and many are out of print. As far as this writer has been able to determine, the only index of these publications is by year of publication. Most of them are of little current value to the superintendent, being either of trade publications at the apprentice level, technical publications at the engineering level, or booklets explaining particular tax and legal matters for the contractor. These legal matters are quite important, but change annually and must be obtained from a current publication.

Minimum Property Standards, the FHA specification book, is essential for any superintendent or builder on FHA-insured construction. It is also valuable for those who want a ready index of standard acceptable construction details without the bulk of an architectural guide. It is well illustrated, and for that reason is more easily understood than is your local building code, which will repeat many of the same requirements. Minimum construction requirements and dimensions are shown, and alternative methods of construction are clearly described. Although trade language is freely used, it is usually the language of the carpenter, although terms used by architects sometimes appear.

Moving the Earth, H. L. Nichols, Jr., 2nd ed., 1962, about 1760 pp., North Castle Books, 212 Bedford Road, Greenwich, Connecticut 06830 ($35.00).

This is a simply written and complete book on excavation. It includes clearing of land, survey methods, types of soils, temporary roads, cellar excavation, ditching, and swamps, with chapters on costs, estimating, and contracts. In short, it is intended as a guide for the excavating contractor, foreman, and operating engineer. (It is consistently advertised in the operators' journal, the *Operating Engineer.*)

Although written for workmen in the construction business, it is not a laymen's book and the new reader must expect to take some time to study it. However, the language of the trade is used, and new words are fully defined and usually illustrated. It has a complete glossary. The author has used extensive information from manufacturers but has not followed the example of other contemporary writers in merely reprinting company information; Nichols appraises his information, and although the book is intended primarily as a description of construction methods and machinery, the author does not hesitate to give his own opinions. Likewise, although there are a number of standard tables and some information that appears in other books on related topics, there is no book in print that offers so much information.

Many chapters are of little interest to building superintendents. Portions on tunneling, earth dams, and "hydraulicking" (moving earth by water nozzle and channels) will rarely be required by the superintendent. However, the simplest excavation problems of buildings are also described, and the building superintendent can usually find a portion of the book that gives him some information on any excavation he encounters and with sufficient detail that he can proceed with his work.

This book is valuable in selecting equipment in that the purposes of equipment are clearly defined and alternative ways are presented of doing the same work—or alternative equipment for the same job. The book is up to date and is the most complete book of its type published in the last 50 years.

The chapter on drilling is particularly helpful to the building superintendent and includes methods of sinking shafts both horizontally and vertically, and of tunneling under roads. Although few people would be expected to read the entire book, it is a reference that will serve you from completion of your apprenticeship to the day you retire from you own construction business, continually serving as a source of information on equipment and methods on any operation involving material that comes out of the ground.

Also available in Spanish.

Superintendents Guide to Theft and Vandalism Prevention, Associated General Contractors of America, 1957 E Street, N.W., Washington, D.C. 20006 ($.50 per copy).

A pamphlet with tips for general superintendents on how to decrease job losses. Simply written and illustrated.

The Use of CPM in Construction, 192 pp., 1976, Associated General Contractors of America, 1957 E Street, N.W., Washington, D.C. 20006 ($10.00).

A rewrite of previous AGC books on the critical-path method (or construction project management), simply written. This is by far the best book on scheduling for construction industry people.

Glossary

A/E. Usually design firms for buildings employ both engineers and architects, and may call themselves architects, engineers, or architect-engineers. Much design work, especially structural design, may be done by either profession. A/E is a common term to include either architectural or engineering firms, or combination of the two.

BANKRUPT. This term is greatly simplified in Chapter 3. For a more detailed explanation, see *The Construction Manager* by this author (Englewood Cliffs, N.J.: Prentice-Hall, Inc., 1974).

BARGAINING UNIT. Both unions and employers organize for uniform contracts in combinations (units) of various sizes. Several unions who contract with the same employer may negotiate together. Both employers and unions may combine a large geographical area and several of their own usually independent units, or they may bargain independently.

BOOKKEEPER. The person who actually makes entries in records, rather than making decisions about how they are to be kept. He is usually supervised by an accountant in the firm or an independent consultant. Many accountants, of course, do their own bookkeeping. A secretary is often a part-time bookkeeper in small offices.

BURNING TORCH. Another name for an acetylene and oxygen cutting torch. It cuts metal by burning it with an oxygen jet—acetylene gas is necessary only to start the operation. A burning torch blows out pieces of melted and burning metal, which are especially likely to start fires. With a change from a burning to a welding torch, the same equipment may be used for acetylene welding, but this is seldom done on construction work.

CARROT AND STICK. Refers to an old story that one makes a donkey go by giving him a carrot in front and hitting him with a stick behind. Donkeys, unlike horses, stop rather arbitrarily when they please.

CARRYING. A foreman may keep an incompetent worker out of friendship or even for payment (kickback). Also, when two craftsmen work together, one may have to direct the other. The latter situation is not important until you must give the workers separate assignments which one of them is incapable of doing.

CHARTER, UNION. A local union obtains a charter from its national headquarters either when formed or when it joins the national organization. This gives the local branch the authority to claim jurisdiction against any other local of the same trade. If the national headquarters revokes the charter, the national union may then establish a rival local, but usually it cannot dissolve an existing local. A charter is similar to a business franchise.

CHECK (verb). A person who examines an invoice, material list, actual amounts of material or work in place, and determines that the item examined corresponds with other records is said to check the item.

COMPUTER PRINTOUT, CPM. If CPM scheduling is used, various kinds of schedules can be obtained, as lists of operations arranged by trade or by dates. The printout is the principal advantage of CPM scheduling.

COST PLUS. A contract arrangement under which the owner pays the actual cost of the work plus a percentage or fixed amount to the contractor. It is used when the work cannot be accurately estimated at the time the contract is made.

CRAFTSMAN. Any skilled worker in construction, usually with particular reference to the quality of work rather than efficiency, supervisory ability, or other attributes.

DEPRECIATION. As used in Chapter 7, this term refers to all fixed costs of equipment, that is, reduction in value of equipment when not working, and interest on amount owed. For practical purposes, it includes interest on equipment owned and payments on equipment not owned. For the purpose of deciding if equipment should work overtime, this use is adequate. It is not the rigid definition used in accounting.

ELASTIC STRUCTURE. Metals are elastic in that they may be compressed or stretched and will return to their original size, as compared with materials such as concrete and wood, which, once out of shape, have little tendency to return. Elastic strings vibrating in harmony create guitar notes, and metal bars give forth xylophone notes.

ENGINEERED DESIGN. A calculation made to determine actual construction requirements, rather than making a selection from what appears reasonable from past experience. This is the application of engineering principles, although it may be done by a master plumber or electrician rather than by an engineer.

EXTENSIONS, TELEPHONE. An instrument on the same line as another, with no telephone central or operator on the site. In offices or large jobs with telephone facilities, an extension is for all practical uses the same as a direct telephone connection.

GAUGE (or GAGE, an alternate spelling). A measurement of thickness of metal sheets and wire. 10-gauge steel is slightly over 1/8 inch thick. Thicker steel is called plate rather than sheet, although 3-gauge sheet steel is nearly 1/4 inch thick. A larger gauge number indicates less thickness; the product of gauge and thickness is about 0.75 to 1.4 in the usual gauges used on construction. Gauge numbers represent sheet weights in multiples of 10 ounces per square foot over 14 gauge, with smaller intervals for lighter material.

GENERAL FOREMAN. When a construction job requires two or more foremen, union rules usually require a general foreman, who is a union member, other than the superintendent. In an open-shop firm, the general foreman may be a superintendent of several associated trades.

GLAZING. The installation of glass, puttying glass, aluminum frames and aluminum frame doors, and similar work is *glazing* and the craftsmen are *glaziers* (not glazers). Glazers are people who apply glaze to pottery. Glazier derives from the name of the trade workers who manufacture glass.

GOOD. In this text, used as in the industry—an efficient worker or supervisor, without an ethical implication.

INSOLVENT. May mean *with no net worth,* or may mean only a shortage of liquid assets (money and property that can be quickly sold) which prevents a firm from paying current debts. The terms "insolvent" and "bankrupt" are greatly simplified in the discussion in Chapter 3.

LEADING MAN. A craftsman designated as an authority over other workers but who continues to work with his tools. He gets higher pay than a craftsman but may not keep time for others or have authority to hire and fire them. Similar to "working foreman" and "pusher."

LIEN LAWS. Allow firms and such persons as craftsmen, surveyors, architects, and contractors to make a claim on the property and construction work for debt, as well as collecting from the firm who hired them on the basis of general assets. If an owner has many debts, the contractor has a better chance to collect the amount due. The bank furnishing the construction financing usually has a claim before the contractor, and this must be paid before the contractor can collect his money. If the owner is a corporation with no other assets than that one job, the chances of collecting defaulting debts are poor.

LIQUIDATED DAMAGES. Deductions from the amount paid the contractor, at a fixed daily rate for late completion of the work, in lieu of other damages. However, since damages not definitely fixed are hard to justify in court, liquidated damages are practically the only kind of damages paid by contractors.

MAGNETIC SIGNS. Thin, flexible plastic signs, which have embedded magnets that cause them to stick to metal surfaces, such as cars and trucks. The magnets are guaranteed for a year but generally last several years. Although easily removable by hand, it is unlikely that they will be dislodged through travel on rough roads.

MORTGAGE INSUROR. The federal government guarantees to a lending institution, usually an insurance company, the payment of mortgages written in accordance with its requirements. If the person who buys the property fails to make the payments, the government pays off the lender and resells the house. Sometimes the loan is made directly by the government agency.

OPEN SHOP. An open-shop contractor is one who is not a signer of a union labor agreement for his own employees. He is also called a merit-shop contractor, to indicate that pay scales are based on skill rather than being uniform. An open-shop contractor often has union subcontractors, and his employees may be union members.

PAY WEEK. Nearly everyone in the construction industry, including executives and office employees, are paid weekly. The pay week is the time for which pay is calculated: for example, beginning on Thursday morning and ending the following Wednesday evening. Payment is made 2 or 3 days after the end of the pay week, as on Friday for a pay week ending on the previous Wednesday.

PRIVATE WORK. Construction for a person or corporation rather than for a government.

PUBLIC WORK. Construction for the federal government; a state, county, or city government; or a government agency.

REQUIREMENTS OF THEIR CRAFT (TOOLS). Each craftsman is obliged to furnish his own hand tools. Very often these are listed in union working rules, along with a statement of the tools union members may not furnish, particularly power tools. Many workers have tools not included in the list, however. The union does not allow power tools to be furnished by members, as these members would be hired before others, and soon all members would be required to furnish power tools.

SEND OUT. A business agent, on request from a job superintendent, sends out the next worker on his list. Under federal rules, the worker is not actually required to be hired until confirmed by the employer; but practically, the employers pay show-up (usually 2 hours' pay) time to the worker if he is not hired. The BA *refers* the member to the employer; the employee carries a slip from the BA to the job superintendent and to the job steward.

STRIKE. A strike is an action by an entire bargaining unit of one union. Usually, other trades do not strike and no pickets are posted. Strikes (as opposed to walkouts) rarely occur except on contract expiration.

TAFT-HARTLEY LAW. The basic federal law regulating what unions may do, and still remain exempt, as all labor unions are, from the anti-monopoly laws to which business firms must conform.

TRADE-OFF. Choice of one course of action which, although costly, is less expensive than another and which has compensating benefits: that is, to balance costs against benefits, for two courses of action.

WALKOUT. A walkout is a local and usually informal strike of a particular union against one employer over a local dispute, usually not over wages. The business agent may or may not authorize the walkout. As no contract negotiations are involved, such actions seldom last more than a few days. They originate over a supposed failure of the employer to conform to the contract, and the business agent is usually helpful in terminating them.

Index